AF539385

GEOMAGNETISM

SOLID EARTH AND UPPER ATMOSPHARE PERSPECTIVES

GEOMAGNETISM
SOLID EARTH AND UPPER ATMOSPHARE PERSPECTIVES

Ravindra Singh

RANDOM PUBLICATIONS

NEW DELHI - 110 002 (INDIA)

Geomagnetism: Solid Earth and Upper Atmosphare Perspectives

ISBN 978-93-51117-06-3

Published in 2015 in India by

RANDOM PUBLICATIONS
4376-A/4B, Gali Murari Lal, Ansari Road
New Delhi-110 002
Phone: +9111-43580356, 23289044
E-mail: randomexports@gmail.com; sales@randompublications.com; info@randompublications.com

Reprinted 2024

Type Setting by: Friends Media, Delhi-110089
Digitally Printed at : Replika Press Pvt. Ltd.

Preface

Earth's magnetic field, also known as the geomagnetic field, is the magnetic field that extends from the Earth's interior to where it meets the solar wind, a stream of charged particles emanating from the Sun. Its magnitude at the Earth's surface ranges from 25 to 65 microtesla (0.25 to 0.65 gauss). Roughly speaking it is the field of a magnetic dipole currently tilted at an angle of about 20 degrees with respect to Earth's rotational axis, as if there were a bar magnet placed at that angle at the centre of the Earth. Unlike a bar magnet, however, Earth's magnetic field changes over time because it is generated by a geodynamo. The North and South magnetic poles wander widely, but sufficiently slowly for ordinary compasses to remain useful for navigation. However, at irregular intervals averaging several hundred thousand years, the Earth's field reverses and the North and South Magnetic Poles relatively abruptly switch places. These reversals of the geomagnetic poles leave a record in rocks that are of value to paleomagnetists in calculating geomagnetic fields in the past. Such information in turn is helpful in studying the motions of continents and ocean floors in the process of plate tectonics.

The magnetosphere is the region above the ionosphere and extends several tens of thousands of kilometres into space, protecting the Earth from the charged particles of the solar wind and cosmic rays that would otherwise strip away the upper atmosphere, including the ozone layer that protects the Earth from harmful ultraviolet radiation. Earth's magnetic field serves to deflect most of the solar wind, whose charged particles would otherwise strip away the ozone layer that protects the Earth from harmful ultraviolet radiation. One stripping mechanism is for gas to be caught in bubbles of magnetic field, which are ripped off by solar winds. Calculations of the loss of carbon dioxide from the atmosphere of Mars, resulting from scavenging of ions by the solar wind, indicate that the dissipation of the magnetic field of Mars caused a near-total loss of its atmosphere. The study of past magnetic field of the Earth is known as paleomagnetism. The polarity of the Earth's magnetic field is recorded in igneous rocks, and reversals of the field are thus detectable as "stripes" centred on mid-ocean ridges where the sea floor is spreading, while the stability of the geomagnetic poles between reversals has allowed paleomagnetists to track the past motion of continents. Reversals also provide the basis for magnetostratigraphy, a way of dating rocks and sediments. The field also magnetizes the crust, and magnetic anomalies can be used to search for deposits of metal ores. Humans have used compasses for direction finding since the 11th century A.D. and for navigation since the 12th century. Although the North Magnetic Pole does shift with time, this wandering is slow enough that a simple compass remains useful for navigation. Using magnetoception various other organisms, ranging from soil bacteria to pigeons, can detect the magnetic field and use it for navigation. Variations in the magnetic field strength have been correlated to rainfall variation within the tropics. The Earth's magnetic field is believed to be generated by electric currents in the conductive material of its core, created by convection currents due to heat escaping from the core. However the process is complex, and computer models that reproduce some of its features have only been developed in the last few decades. The North and South magnetic poles wander widely, but sufficiently slowly for ordinary compasses to remain useful for navigation. However, at irregular intervals averaging several hundred thousand years, the Earth's field reverses and the North and South Magnetic Poles relatively abruptly switch places.

The main magnetic field of the Earth is a complex physical phenomenon. This book presents in detail the foundations of geomagnetism, developed from first principles. Several important aspects of solid Earth geomagnetism are elaborated in the book.

I thank all members of my team who have helped in the preparation of the book. My special thanks go to "Random Publications" who have published the book.

— Ravindra Singh

Contents

Chapter 1

Introduction

Earth's magnetic field, also known as the geomagnetic field, is the magnetic field that extends from the Earth's interior to where it meets the solar wind, a stream of charged particles emanating from the Sun. Its magnitude at the Earth's surface ranges from 25 to 65 microtesla (0.25 to 0.65 gauss). Roughly speaking it is the field of a magnetic dipole currently tilted at an angle of about 20 degrees with respect to Earth's rotational axis, as if there were a bar magnet placed at that angle at the center of the Earth. Unlike a bar magnet, however, Earth's magnetic field changes over time because it is generated by a geodynamo (in Earth's case, the motion of molten iron alloys in its outer core).

The North and South magnetic poles wander widely, but sufficiently slowly for ordinary compasses to remain useful for navigation. However, at irregular intervals averaging several hundred thousand years, the Earth's field reverses and the North and South Magnetic Poles relatively abruptly switch places. These reversals of the geomagnetic poles leave a record in rocks that are of value to paleomagnetists in calculating geomagnetic fields in the past. Such information in turn is helpful in studying the motions of continents and ocean floors in the process of plate tectonics.

The magnetosphere is the region above the ionosphere and extends several tens of thousands of kilometers into space, protecting the Earth from the charged particles of the solar wind and cosmic rays that would otherwise strip away the upper atmosphere, including the ozone layer that protects the Earth from harmful ultraviolet radiation.

The study of geomagnetism is one of the oldest of the geophysical sciences. The geomagnetic fields have been observed and used from ancient times. Modern uses of geomagnetic data include navigation and mineral exploration.

Importance

Earth's magnetic field serves to deflect most of the solar wind, whose charged particles would otherwise strip away the ozone layer that protects the Earth from harmful ultraviolet radiation. One stripping mechanism is for gas to be caught in bubbles of magnetic field, which are ripped off by solar winds. Calculations of the loss of carbon dioxide from the atmosphere of Mars, resulting from scavenging of ions by the solar wind, indicate that the dissipation of the magnetic field of Mars caused a near-total loss of its atmosphere.

The study of past magnetic field of the Earth is known as paleomagnetism. The polarity of the Earth's magnetic field is recorded in igneous rocks, and reversals of the field are thus detectable as "stripes" centered on mid-ocean ridges where the sea floor is spreading, while the stability of the geomagnetic poles between reversals has allowed paleomagnetists to track the past motion of continents. Reversals also provide the basis for magnetostratigraphy, a way of dating rocks and sediments. The field also magnetizes the crust, and magnetic anomalies can be used to search for deposits of metal ores.

Humans have used compasses for direction finding since the 11th century A.D. and for navigation since the 12th century. Although the North Magnetic Pole does shift with time, this wandering is slow enough that a simple compass remains useful for navigation. Using magnetoception various other organisms, ranging from soil bacteria to pigeons, can detect the magnetic field and use it for navigation.

Variations in the magnetic field strength have been correlated to rainfall variation within the tropics.

Main Characteristics

Description

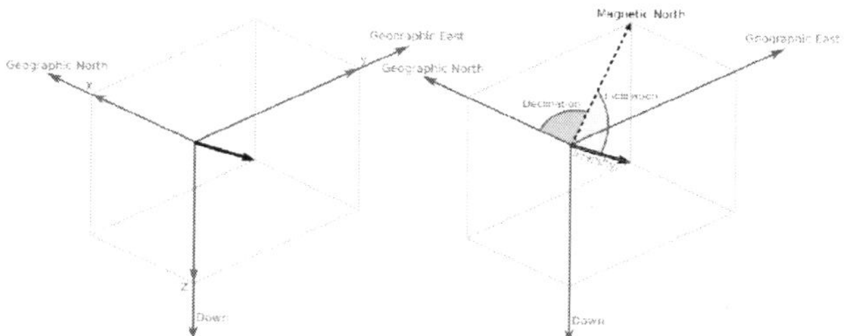

Figure: *Common coordinate systems used for representing the Earth's magnetic field.*

At any location, the Earth's magnetic field can be represented by a three-dimensional vector. A typical procedure for measuring its direction is to use a compass to determine the direction of magnetic North. Its angle relative to true North is the *declination* (*D*) or *variation*. Facing magnetic North, the angle the field makes with the horizontal is the *inclination* (*I*) or *dip*. The *intensity* (*F*) of the field is proportional to the force it exerts on a magnet. Another common representation is in *X* (North), *Y* (East) and *Z* (Down) coordinates.

Intensity

The intensity of the field is often measured in gauss (G), but is generally reported in nanotesla (nT), with 1 G = 100,000 nT. A nanotesla is also referred to as a gamma (γ). The tesla is the SI unit of the Magnetic field, B. The field ranges between approximately 25,000 and 65,000 nT (0.25–0.65 G). By comparison, a strong refrigerator magnet has a field of about 100 gauss (0.010 T).

A map of intensity contours is called an *isodynamic chart*. As the 2010 World Magnetic Model shows, the intensity tends to decrease from the poles to the equator. A minimum intensity occurs over South America while there are maxima over northern Canada, Siberia, and the coast of Antarctica south of Australia.

Inclination

The inclination is given by an angle that can assume values between -90° (up) to 50° (down). In the northern hemisphere, the field points downwards. It is straight down at the North Magnetic Pole and rotates upwards as the latitude decreases until it is horizontal (0°) at the magnetic equator. It continues to rotate upwards until it is straight up at the South Magnetic Pole. Inclination can be measured with a dip circle.

An *isoclinic chart* (map of inclination contours) for the Earth's magnetic field is shown below.

Declination

Declination is positive for an eastward deviation of the field relative to true north. It can be estimated by comparing the magnetic north/ south heading on a compass with the direction of a celestial pole. Maps typically include information on the declination as an angle or a small diagram showing the relationship between magnetic north and true north. Information on declination for a region can be represented by a chart with isogonic lines (contour lines with each line representing a fixed declination).

Geographical Variation

Components of the Earth's magnetic field at the surface from the World Magnetic Model for 2010.

Dipolar Approximation

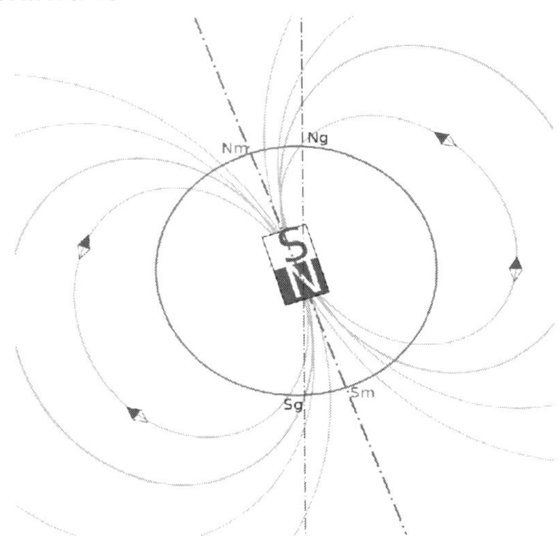

Figure: *The variation between magnetic north (N_m) and "true" north (N_g).*

Near the surface of the Earth, its magnetic field can be closely approximated by the field of a magnetic dipole positioned at the center of the Earth and tilted at an angle of about 10° with respect to the rotational axis of the Earth. The dipole is roughly equivalent to a powerful bar magnet, with its south pole pointing towards the geomagnetic North Pole. This may seem surprising, but the north pole of a magnet is so defined because, if allowed to rotate freely, it points roughly northward (in the geographic sense). Since the north pole of a magnet attracts the south poles of other magnets and repels the north poles, it must be attracted to the south pole of Earth's magnet. The dipolar field accounts for 80–90% of the field in most locations.

Magnetic Poles

The positions of the magnetic poles can be defined in at least two ways: locally or globally.

One way to define a pole is as a point where the magnetic field is vertical. This can be determined by measuring the inclination, as described above. The inclination of the Earth's field is 50° (upwards) at the North Magnetic Pole and -90°(downwards) at the South Magnetic Pole. The two poles wander independently of each other and are not directly opposite each other on the globe. They can migrate rapidly:

movements of up to 40 kilometres (25 mi) per year have been observed for the North Magnetic Pole. Over the last 180 years, the North Magnetic Pole has been migrating northwestward, from Cape Adelaide in the Boothia Peninsula in 1831 to 600 kilometres (370 mi) from Resolute Bay in 2001. The *magnetic equator* is the line where the inclination is zero (the magnetic field is horizontal).

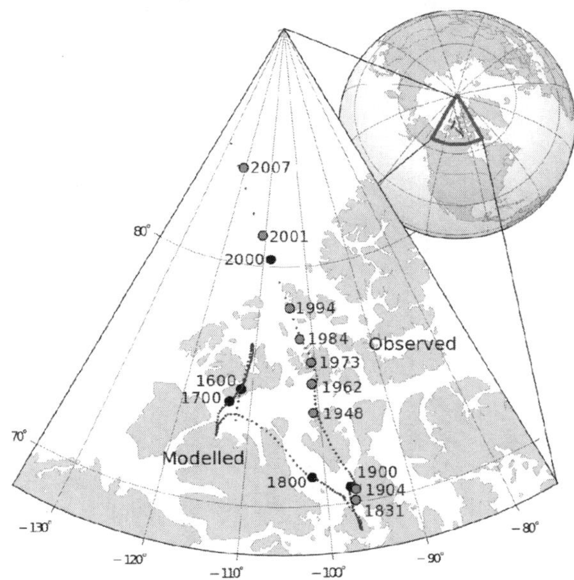

Figure: *The movement of Earth's North Magnetic Pole across the Canadian arctic, 1831–2007.*

The global definition of the Earth's field is based on a mathematical model. If a line is drawn through the center of the Earth, parallel to the moment of the best-fitting magnetic dipole, the two positions where it intersects the Earth's surface are called the North and South geomagnetic poles. If the Earth's magnetic field were perfectly dipolar, the geomagnetic poles and magnetic dip poles would coincide and compasses would point towards them. However, the Earth's field has a significant non-dipolar contribution, so the poles do not coincide and compasses do not generally point at either.

Magnetosphere

Earth's magnetic field, predominantly dipolar at its surface, is distorted further out by the solar wind. This is a stream of charged particles leaving the Sun's corona and accelerating to a speed of 200 to 1000 kilometres per second. They carry with them a magnetic field, the interplanetary magnetic field (IMF).

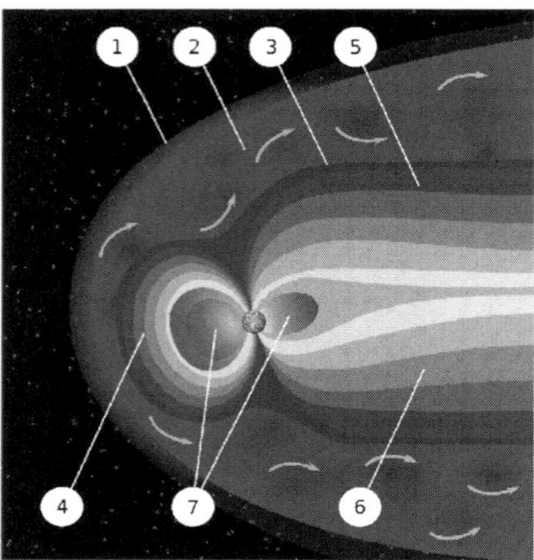

Figure: *An artist's rendering of the structure of a magnetosphere. 1) Bow shock. 2) Magnetosheath. 3) Magnetopause. 4) Magnetosphere. 5) Northern tail lobe. 6) Southern tail lobe. 7) Plasmasphere.*

The solar wind exerts a pressure, and if it could reach Earth's atmosphere it would erode it. However, it is kept away by the pressure of the Earth's magnetic field. The magnetopause, the area where the pressures balance, is the boundary of the magnetosphere. Despite its name, the magnetosphere is asymmetric, with the sunward side being about 10 Earth radii out but the other side stretching out in a magnetotail that extends beyond 200 Earth radii. Sunward of the magnetopause is the bow shock, the area where the solar wind slows abruptly.

Inside the magnetosphere is the plasmasphere, a donut-shaped region containing low-energy charged particles, or plasma. This region begins at a height of 60 km, extends up to 3 or 4 Earth radii, and includes the ionosphere.

This region rotates with the Earth. There are also two concentric tire-shaped regions, called the Van Allen radiation belts, with high-energy ions (energies from 0.1 to 10 million electron volts (MeV)). The inner belt is 1–2 Earth radii out while the outer belt is at 4–7 Earth radii. The plasmasphere and Van Allen belts have partial overlap, with the extent of overlap varying greatly with solar activity.

As well as deflecting the solar wind, the Earth's magnetic field deflects cosmic rays, high-energy charged particles that are mostly from outside the Solar system. (Many cosmic rays are kept out of the Solar system by the Sun's magnetosphere, or heliosphere.) By contrast, astronauts on the Moon risk exposure to radiation. Anyone who had been on the Moon's surface during a particularly violent solar eruption in 2005 would have received a lethal dose.

Some of the charged particles do get into the magnetosphere. These spiral around field lines, bouncing back and forth between the poles several times per second. In addition, positive ions slowly drift westward and negative ions drift eastward, giving rise to a ring current. This current reduces the magnetic field at the Earth's surface. Particles that penetrate the ionosphere and collide with the atoms there give rise to the lights of the aurorae and also emit X-rays.

The varying conditions in the magnetosphere, known as space weather, are largely driven by solar activity. If the solar wind is weak, the magnetosphere expands; while if it is strong, it compresses the magnetosphere and more of it gets in. Periods of particularly intense activity, called geomagnetic storms, can occur when a coronal mass ejection erupts above the Sun and sends a shock wave through the Solar System.

Such a wave can take just two days to reach the Earth. Geomagnetic storms can cause a lot of disruption; the "Halloween" storm of 2003 damaged more than a third of NASA's satellites. The largest documented storm occurred in 1859. It induced currents strong enough to short out telegraph lines, and aurorae were reported as far south as Hawaii.

Time Dependence

Short-Term Variations

The geomagnetic field changes on time scales from milliseconds to millions of years. Shorter time scales mostly arise from currents in the ionosphere (ionospheric dynamo region) and magnetosphere, and some changes can be traced to geomagnetic storms or daily variations in currents. Changes over time scales of a year or more mostly reflect changes in the Earth's interior, particularly the iron-rich core.

Frequently, the Earth's magnetosphere is hit by solar flares causing geomagnetic storms, provoking displays of aurorae. The short-term instability of the magnetic field is measured with the K-index.

Data from THEMIS show that the magnetic field, which interacts with the solar wind, is reduced when the magnetic orientation is aligned between Sun and Earth - opposite to the previous hypothesis. During forthcoming solar storms, this could result in blackouts and disruptions in artificial satellites.

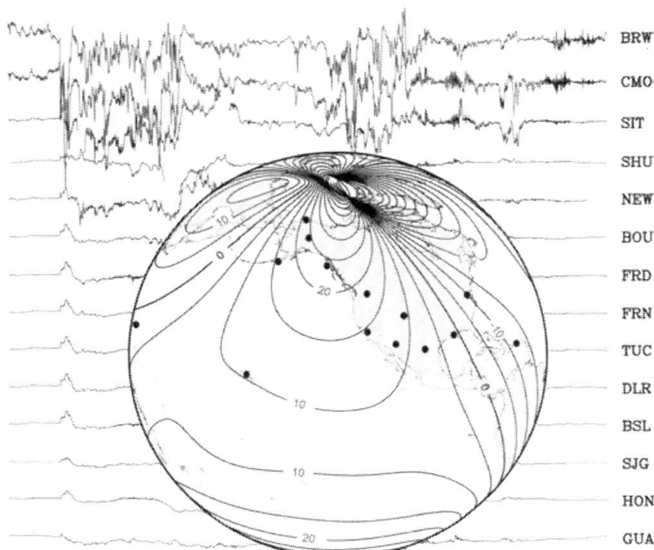

Figure: *Background: a set of traces from magnetic observatories showing a magnetic storm in 2000.*
Globe: map showing locations of observatories and contour lines giving horizontal magnetic intensity in μ *T.*

Secular Variation

Changes in Earth's magnetic field on a time scale of a year or more are referred to as *secular variation.* Over hundreds of years, magnetic declination is observed to vary over tens of degrees. A movie on the right shows how global declinations have changed over the last few centuries.

The direction and intensity of the dipole change over time. Over the last two centuries the dipole strength has been decreasing at a rate of about 6.3% per century. At this rate of decrease, the field would be negligible in about 1600 years. However, this strength is about average for the last 7 thousand years, and the current rate of change is not unusual.

A prominent feature in the non-dipolar part of the secular variation is a *westward drift* at a rate of about 0.2 degrees per year. This drift is not the same everywhere and has varied over time. The globally averaged drift has been westward since about 1400 AD but eastward between about 1000 AD and 1400 AD.

Changes that predate magnetic observatories are recorded in archaeological and geological materials. Such changes are referred to as *paleomagnetic secular variation* or *paleosecular variation (PSV)*. The records typically include long periods of small change with occasional large changes reflecting geomagnetic excursions and reversals.

Magnetic Field Reversals

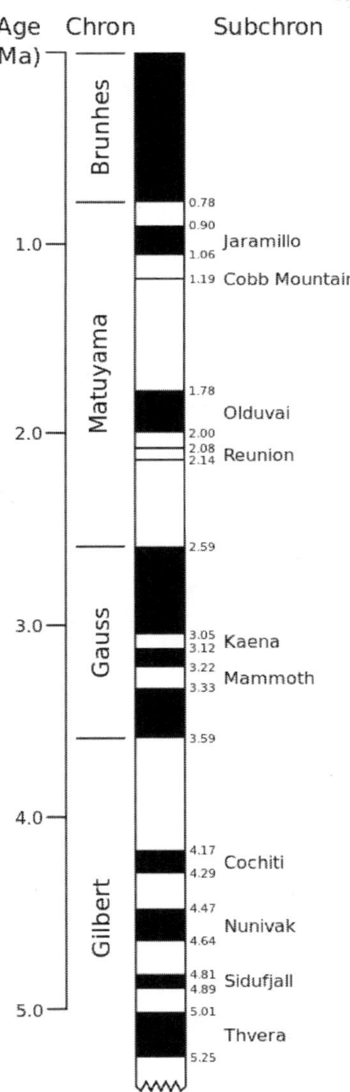

Figure: *Geomagnetic polarity during the late Cenozoic Era. Dark areas denote periods where the polarity matches today's polarity, light areas denote periods where that polarity is reversed.*

Although the Earth's field is generally well approximated by a magnetic dipole with its axis near the rotational axis, there are occasional dramatic events where the North and South geomagnetic poles trade places. Evidence for these *geomagnetic reversals* can be found worldwide in basalts, sediment cores taken from the ocean floors, and seafloor magnetic anomalies. Reversals occur at apparently random intervals ranging from less than 0.1 million years to as much as 50 million years. The most recent geomagnetic reversal, called the Brunhes–Matuyama reversal, occurred about 780,000 years ago. Another global reversal of the Earth's field, called the Laschamp event, occurred during the last ice age (41,000 years ago). However, because of its brief duration it is labelled an *excursion*.

The past magnetic field is recorded mostly by iron oxides, such as magnetite, that have some form of ferrimagnetism or other magnetic ordering that allows the Earth's field to magnetize them. This remanent magnetization, or *remanence*, can be acquired in more than one way. In lava flows, the direction of the field is "frozen" in small magnetic particles as they cool, giving rise to a thermoremanent magnetization. In sediments, the orientation of magnetic particles acquires a slight bias towards the magnetic field as they are deposited on an ocean floor or lake bottom. This is called *detrital remanent magnetization*.

Thermoremanent magnetization is the form of remanence that gives rise to the magnetic anomalies around ocean ridges. As the seafloor spreads, magma wells up from the mantle and cools to form new basaltic crust. During the cooling, the basalt records the direction of the Earth's field. This new basalt forms on both sides of the ridge and moves away from it. When the Earth's field reverses, new basalt records the reversed direction. The result is a series of stripes that are symmetric about the ridge. A ship towing a magnetometer on the surface of the ocean can detect these stripes and infer the age of the ocean floor below. This provides information on the rate at which seafloor has spread in the past.

Radiometric dating of lava flows has been used to establish a *geomagnetic polarity time scale*, part of which is shown in the image. This forms the basis of magnetostratigraphy, a geophysical correlation technique that can be used to date both sedimentary and volcanic sequences as well as the seafloor magnetic anomalies.

Studies of lava flows on Steens Mountain, Oregon, indicate that the magnetic field could have shifted at a rate of up to 6 degrees per day at some time in Earth's history, which significantly challenges the popular understanding of how the Earth's magnetic field works.

Temporary dipole tilt variations that take the dipole axis across the equator and then back to the original polarity are known as *excursions*.

Earliest Appearance

A paleomagnetic study of Australian red dacite and pillow basalt has estimated the magnetic field to have been present since at least 3,450 million years ago.

Future

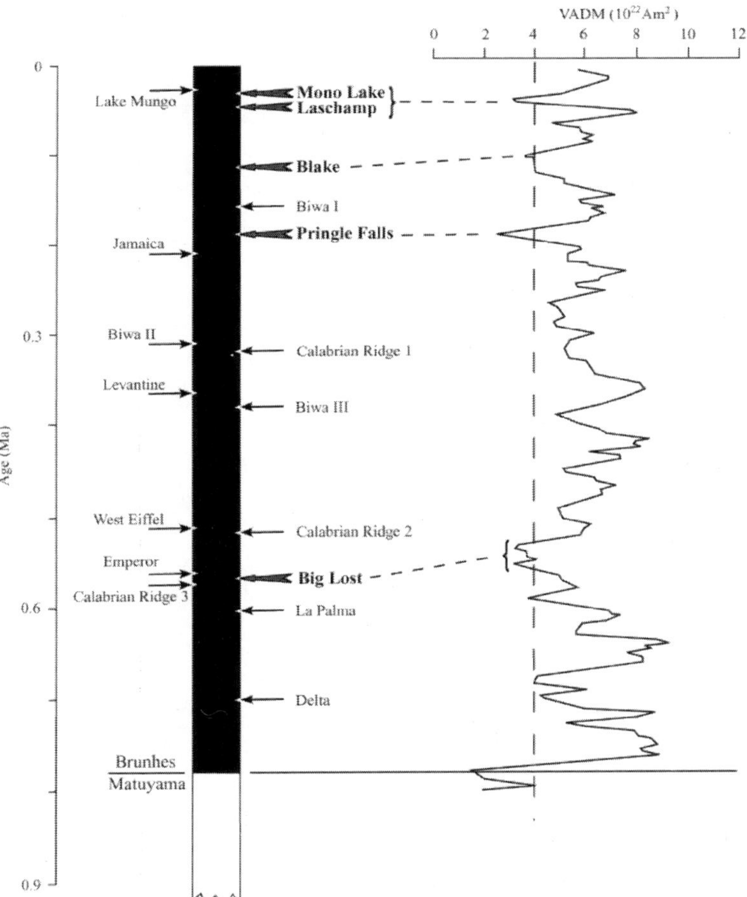

Figure: *Variations in virtual axial dipole moment since the last reversal.*

At present, the overall geomagnetic field is becoming weaker; the present strong deterioration corresponds to a 10–15% decline over the last 150 years and has accelerated in the past several years; geomagnetic intensity has declined almost continuously from a maximum 35% above the modern value achieved approximately 2,000 years ago. The rate of decrease and the current strength are within the normal range of variation, as shown by the record of past magnetic fields recorded in rocks.

The nature of Earth's magnetic field is one of heteroscedastic fluctuation. An instantaneous measurement of it, or several measurements of it across the span of decades or centuries, are not sufficient to extrapolate an overall trend in the field strength. It has gone up and down in the past for no apparent reason. Also, noting the local intensity of the dipole field (or its fluctuation) is insufficient to characterize Earth's magnetic field as a whole, as it is not strictly a dipole field. The dipole component of Earth's field can diminish even while the total magnetic field remains the same or increases.

The Earth's magnetic north pole is drifting from northern Canada towards Siberia with a presently accelerating rate—10 kilometres (6.2 mi) per year at the beginning of the 20th century, up to 40 kilometres (25 mi) per year in 2003, and since then has only accelerated.

Physical Origin

The Earth's magnetic field is believed to be generated by electric currents in the conductive material of its core, created by convection currents due to heat escaping from the core. However the process is complex, and computer models that reproduce some of its features have only been developed in the last few decades.

Earth's Core and the Geodynamo

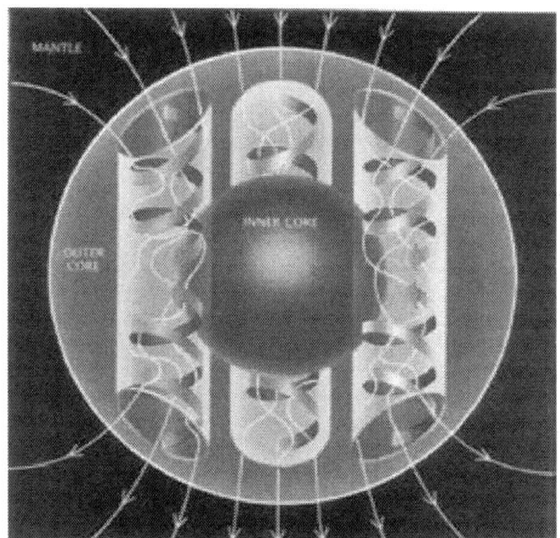

Figure: *A schematic illustrating the relationship between motion of conducting fluid, organised into rolls by the Coriolis force, and the magnetic field the motion generates.*

The Earth and most of the planets in the Solar System, as well as the Sun and other stars, all generate magnetic fields through the motion of highly conductive fluids. The Earth's field originates in its core. This is a region of iron alloys extending to about 3400 km (the radius of the Earth is 6370 km). It is divided into a solid inner core, with a radius of 1220 km, and a liquid outer core. The motion of the liquid in the outer core is driven by heat flow from the inner core, which is about 6,000 K (5,730 °C; 10,340 °F), to the core-mantle boundary, which is about 3,800 K (3,530 °C; 6,380 °F). The pattern of flow is organised by the rotation of the Earth and the presence of the solid inner core.

The mechanism by which the Earth generates a magnetic field is known as a dynamo. A magnetic field is generated by a feedback loop: current loops generate magnetic fields (Ampθre's circuital law); a changing magnetic field generates an electric field (Faraday's law); and the electric and magnetic fields exert a force on the charges that are flowing in currents (the Lorentz force). These effects can be combined in a partial differential equation for the magnetic field called the *magnetic induction equation*:

$$\frac{\partial B}{\partial t} = \eta \nabla^2 B + \nabla \times (u \times B)$$

...where u is the velocity of the fluid; B is the magnetic B-field; and $\eta = 1/\sigma\mu$ is the magnetic diffusivity, a product of the electrical conductivity σ and the permeability μ . The term "B/"t is the time derivative of the field; ∇^2 is the Laplace operator and $\nabla \times$ is the curl operator.

The first term on the right hand side of the induction equation is a diffusion term. In a stationary fluid, the magnetic field declines and any concentrations of field spread out. If the Earth's dynamo shut off, the dipole part would disappear in a few tens of thousands of years.

In a perfect conductor ($\sigma = \infty$), there would be no diffusion. By Lenz's law, any change in the magnetic field would be immediately opposed by currents, so the flux through a given volume of fluid could not change. As the fluid moved, the magnetic field would go with it. The theorem describing this effect is called the *frozen-in-field theorem*. Even in a fluid with a finite conductivity, new field is generated by stretching field lines as the fluid moves in ways that deform it. This process could go on generating new field indefinitely, were it not that as the magnetic field increases in strength, it resists fluid motion.

The motion of the fluid is sustained by convection, motion driven by buoyancy. The temperature increases towards the center of the

Earth, and the higher temperature of the fluid lower down makes it buoyant. This buoyancy is enhanced by chemical separation: As the core cools, some of the molten iron solidifies and is plated to the inner core. In the process, lighter elements are left behind in the fluid, making it lighter. This is called *compositional convection*. A Coriolis effect, caused by the overall planetary rotation, tends to organise the flow into rolls aligned along the north-south polar axis.

The average magnetic field in the Earth's outer core was calculated to be 25 gauss, 50 times stronger than the field at the surface.

Numerical Models

Simulating the geodynamo requires numerically solving a set of nonlinear partial differential equations for the magnetohydrodynamics (MHD) of the Earth's interior. Simulation of the MHD equations is performed on a 3D grid of points and the fineness of the grid, which in part determines the realism of the solutions, is limited mainly by computer power. For decades, theorists were confined to creating *kinematic dynamos* in which the fluid motion is chosen in advance and the effect on the magnetic field calculated. Kinematic dynamo theory was mainly a matter of trying different flow geometries and testing whether such geometries could sustain a dynamo.

The first *self-consistent* dynamo models, ones that determine both the fluid motions and the magnetic field, were developed by two groups in 1995, one in Japan and one in the United States. The latter received attention because it successfully reproduced some of the characteristics of the Earth's field, including geomagnetic reversals.

Currents in the Ionosphere and Magnetosphere

Electric currents induced in the ionosphere generate magnetic fields (ionospheric dynamo region). Such a field is always generated near where the atmosphere is closest to the Sun, causing daily alterations that can deflect surface magnetic fields by as much as one degree. Typical daily variations of field strength are about 25 nanoteslas (nT) (one part in 2000), with variations over a few seconds of typically around 1 nT (one part in 50,000).

Measurement and Analysis

Detection

The Earth's magnetic field strength was measured by Carl Friedrich Gauss in 1835 and has been repeatedly measured since then, showing a relative decay of about 10% over the last 150 years. The

Magsat satellite and later satellites have used 3-axis vector magnetometers to probe the 3-D structure of the Earth's magnetic field. The later Ørsted satellite allowed a comparison indicating a dynamic geodynamo in action that appears to be giving rise to an alternate pole under the Atlantic Ocean west of S. Africa.

Governments sometimes operate units that specialise in measurement of the Earth's magnetic field. These are geomagnetic observatories, typically part of a national Geological survey, for example the British Geological Survey's Eskdalemuir Observatory. Such observatories can measure and forecast magnetic conditions such as magnetic storms that sometimes affect communications, electric power, and other human activities.

The International Real-time Magnetic Observatory Network, with over 100 interlinked geomagnetic observatories around the world has been recording the earths magnetic field since 1991.

The military determines local geomagnetic field characteristics, in order to detect *anomalies* in the natural background that might be caused by a significant metallic object such as a submerged submarine. Typically, these magnetic anomaly detectors are flown in aircraft like the UK's Nimrod or towed as an instrument or an array of instruments from surface ships.

Commercially, geophysical prospecting companies also use magnetic detectors to identify naturally occurring anomalies from ore bodies, such as the Kursk Magnetic Anomaly.

Crustal Magnetic Anomalies

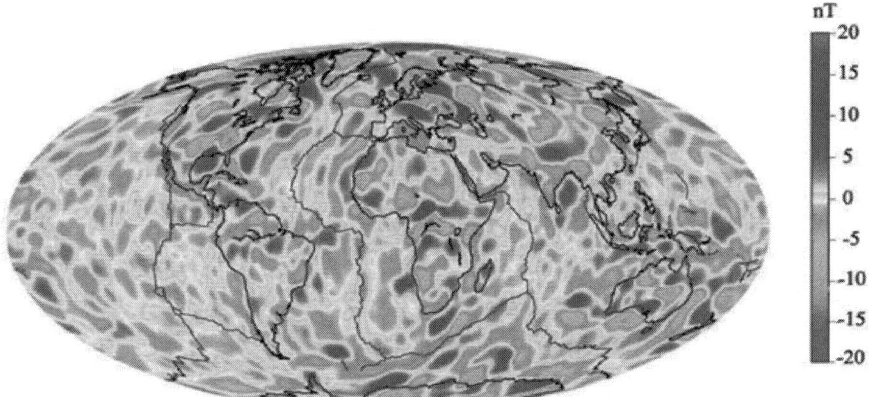

Figure: *A model of short-wavelength features of Earth's magnetic field, attributed to lithospheric anomalies.*

Magnetometers detect minute deviations in the Earth's magnetic field caused by iron artifacts, kilns, some types of stone structures, and even ditches and middens in archaeological geophysics. Using magnetic instruments adapted from airborne magnetic anomaly detectors developed during World War II to detect submarines, the magnetic variations across the ocean floor have been mapped. Basalt — the iron-rich, volcanic rock making up the ocean floor — contains a strongly magnetic mineral (magnetite) and can locally distort compass readings. The distortion was recognised by Icelandic mariners as early as the late 18th century. More important, because the presence of magnetite gives the basalt measurable magnetic properties, these magnetic variations have provided another means to study the deep ocean floor. When newly formed rock cools, such magnetic materials record the Earth's magnetic field.

Statistical Models

Each measurement of the magnetic field is at a particular place and time. If an accurate estimate of the field at some other place and time is needed, the measurements must be converted to a model and the model used to make predictions.

Spherical Harmonics

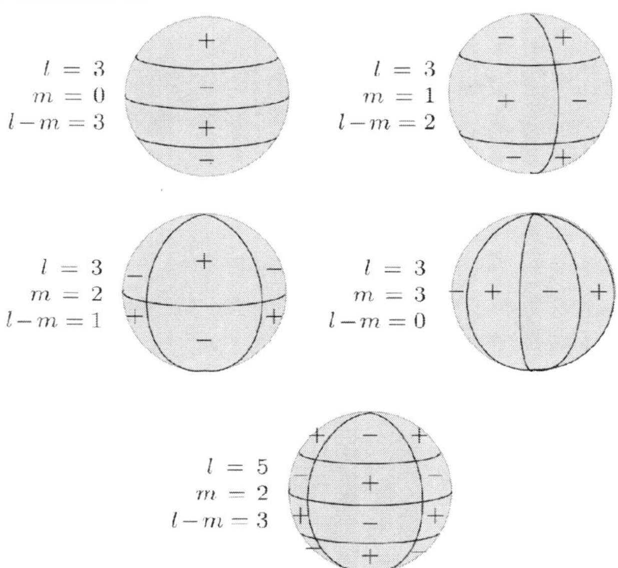

Figure: *Schematic representation of spherical harmonics on a sphere and their nodal lines.* $P_{\ell m}$ *is equal to* ***0*** *along m great circles passing through the poles, and along* $\ell - m$ *circles of equal latitude. The function changes sign each –!time it crosses one of these lines.*

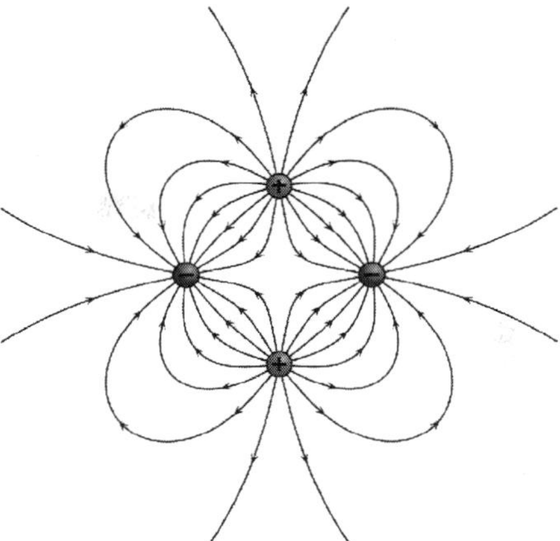

Figure: *Example of a quadrupole field. This could also be constructed by moving two dipoles together. If this arrangement were placed at the center of the Earth, then a magnetic survey at the surface would find two magnetic north poles (at the geographic poles) and two south poles at the equator.*

The most common way of analyzing the global variations in the Earth's magnetic field is to fit the measurements to a set of spherical harmonics. This was first done by Carl Friedrich Gauss. Spherical harmonics are functions that oscillate over the surface of a sphere. They are the product of two functions, one that depends on latitude and one on longitude. The function of longitude is zero along zero or more great circles passing through the North and South Poles; the number of such *nodal lines* is the absolute value of the *order m*. The function of latitude is zero along zero or more latitude circles; this plus the order is equal to the *degree* ℓ. Each harmonic is equivalent to a particular arrangement of magnetic charges at the center of the Earth. A *monopole* is an isolated magnetic charge, which has never been observed. A *dipole* is equivalent to two opposing charges brought close together and a *quadrupole* to two dipoles brought together. A quadrupole field is shown in the lower figure on the right.

Spherical harmonics can represent any scalar field (function of position) that satisfies certain properties. A magnetic field is a vector field, but if it is expressed in Cartesian components *X, Y, Z,* each component is the derivative of the same scalar function called the *magnetic potential.* Analyses of the Earth's magnetic field use a modified version of the usual spherical harmonics that differ by a multiplicative factor. A least-squares fit to the magnetic field measurements gives the Earth's field as the sum of spherical harmonics, each multiplied by the best-fitting *Gauss coefficient* g_m^ℓ or h_m^ℓ.

The lowest-degree Gauss coefficient, g_0^0, gives the contribution of an isolated magnetic charge, so it is zero. The next three coefficients – g_1^0, g_1^1, and h_1^1 – determine the direction and magnitude of the dipole contribution. The best fitting dipole is tilted at an angle of about 10° with respect to the rotational axis, as described earlier.

Radial Dependence

Spherical harmonic analysis can be used to distinguish internal from external sources if measurements are available at more than one height (for example, ground observatories and satellites). In that case, each term with coefficient g_m^ℓ or h_m^ℓ can be split into two terms: one that decreases with radius as $1/r^{\ell+1}$ and one that *increases* with radius as r^ℓ. The increasing terms fit the external sources (currents in the ionosphere and magnetosphere). However, averaged over a few years the external contributions average to zero.

The remaining terms predict that the potential of a dipole source $(\ell = 1)$ drops off as $1/r^2$. The magnetic field, being a derivative of the potential, drops off as $1/r^3$. Quadrupole terms drop off as $1/r^4$, and higher order terms drop off increasingly rapidly with the radius. The radius of the outer core is about half of the radius of the Earth. If the field at the core-mantle boundary is fit to spherical harmonics, the dipole part is smaller by a factor of about 8 at the surface, the quadrupole part by a factor of 16, and so on. Thus, only the components with large wavelengths can be noticeable at the surface. From a variety of arguments, it is usually assumed that only terms up to degree 14 or less have their origin in the core. These have wavelengths of about 2,000 kilometres (1,200 mi) or less. Smaller features are attributed to crustal anomalies.

Global Models

The International Association of Geomagnetism and Aeronomy maintains a standard global field model called the International Geomagnetic Reference Field. It is updated every 5 years. The 11th-generation model, IGRF11, was developed using data from satellites (Ørsted, CHAMP and SAC-C) and a world network of geomagnetic observatories. The spherical harmonic expansion was truncated at degree 10, with 120 coefficients, until 2000. Subsequent models are truncated at degree 13 (195 coefficients).

Another global field model, called World Magnetic Model, is produced jointly by the National Geophysical Data Center and the

British Geological Survey. This model truncates at degree 12 (168 coefficients). It is the model used by the United States Department of Defence, the Ministry of Defence (United Kingdom), the North Atlantic Treaty Organisation, and the International Hydrographic Office as well as in many civilian navigation systems.

A third model, produced by the Goddard Space Flight Center (NASA and GSFC) and the Danish Space Research Institute, uses a "comprehensive modelling" approach that attempts to reconcile data with greatly varying temporal and spatial resolution from ground and satellite sources.

Biomagnetism

Animals including birds and turtles can detect the Earth's magnetic field, and use the field to navigate during migration. Cows and wild deer tend to align their bodies north-south while relaxing, but not when the animals are under high voltage power lines, leading researchers to believe magnetism is responsible. In 2011 a group of Czech researchers reported their failed attempt to replicate the finding using different Google Earth images.

Multipole Expansion

A multipole expansion is a mathematical series representing a function that depends on angles — usually the two angles on a sphere. These series are useful because they can often be truncated, meaning that only the first few terms need to be retained for a good approximation to the original function.

The function being expanded may be complex in general. Multipole expansions are very frequently used in the study of electromagnetic and gravitational fields, where the fields at distant points are given in terms of sources in a small region. The multipole expansion with angles is often combined with an expansion in radius. Such a combination gives an expansion describing a function throughout three-dimensional space.

The multipole expansion is expressed as a sum of terms with progressively finer angular features. For example, the initial term — called the zero-th, or monopole, moment — is a constant, independent of angle. The following term — the first, or dipole, moment — varies once from positive to negative around the sphere. Higher-order terms (like the quadrupole and octupole) vary more quickly with angles. A multipole moment usually involves powers (or inverse powers) of the distance to the origin, as well as some angular dependence.

In principle, a multipole expansion provides an exact description of the potential and generally converges under two conditions:

(1) if the sources (e.g., charges) are localized close to the origin and the point at which the potential is observed is far from the origin; or

(2) the reverse, i.e., if the sources (e.g., charges) are located far from the origin and the potential is observed close to the origin.

In the first (more common) case, the coefficients of the series expansion are called *exterior multipole moments* or simply *multipole moments* whereas, in the second case, they are called *interior multipole moments*. The zeroth-order term in the expansion is called the monopole moment, the first-order term is denoted as the dipole moment, and the third (the second-order), fourth (the third-order), etc. terms are denoted as quadrupole, octupole, etc. moments.

Expansion in Spherical Harmonics

Most commonly, the series is written as a sum of spherical harmonics. Thus, we might write a function $f(\theta,\phi)$ as the sum

$$f(\theta,\phi)=\sum_{l=0}^{\infty}\sum_{m=-l}^{l} C_l^m Y_l^m(\theta,\phi).$$

Here, $Y_l^m(\theta,\phi)$ are the standard spherical harmonics, and C_l^m are constant coefficients which depend on the function. The term C_0^0 represents the monopole; C_1^{-1}, C_1^0, C_1^1 represent the dipole; and so on. Equivalently, the series is also frequently written as

$$f(\theta,\phi)=C+C_i n^i+C_{ij}n^i n^j+C_{ijk}n^i n^j n^k+C_{ijkl}n^i n^j n^k n^l+\cdots.$$

Here, the n^i represent the components of a unit vector in the direction given by the angles θ and ϕ, and indices are implicitly summed. Here, the term C is the monopole; C_i is a set of three numbers representing the dipole; and so on.

In the above expansions, the coefficients may be real or complex. If the function being expressed as a multipole expansion is real, however, the coefficients must satisfy certain properties. In the spherical harmonic expansion, we must have

$$C_l^m=(-1)^m C_l^{m*}.$$

In the multi-vector expansion, each coefficient must be real:

$$C=C^*;\ C_i=C_i^*;\ C_{ij}=C_{ij}^*;\ C_{ijk}=C_{ijk}^*;\ \ldots$$

While expansions of scalar functions are by far the most common application of multipole expansions, they may also be generalized to describe tensors of arbitrary rank. This finds use in multipole expansions of the vector potential in electromagnetism, or the metric perturbation in the description of gravitational waves.

For describing functions of three dimensions, away from the coordinate origin, the coefficients of the multipole expansion can be written as functions of the distance to the origin, r —most frequently, as a Laurent series in powers of r. For example, to describe the electromagnetic potential, V, from a source in a small region near the origin, the coefficients may be written as:

$$V(r,\theta,\phi)=\sum_{l=0}^{\infty}\sum_{m=-l}^{l}C_l^m(r)Y_l^m(\theta,\phi)=\sum_{j=1}^{\infty}\sum_{l=0}^{\infty}\sum_{m=-l}^{l}\frac{D_{l,j}^m}{r^j}Y_l^m(\theta,\phi).$$

Applications of Multipole Expansions

Multipole expansions are widely used in problems involving gravitational fields of systems of masses, electric and magnetic fields of charge and current distributions, and the propagation of electromagnetic waves. A classic example is the calculation of the *exterior* multipole moments of atomic nuclei from their interaction energies with the *interior* multipoles of the electronic orbitals. The multipole moments of the nuclei report on the distribution of charges within the nucleus and, thus, on the shape of the nucleus. Truncation of the multipole expansion to its first non-zero term is often useful for theoretical calculations.

Multipole expansions are also useful in numerical simulations, and form the basis of the Fast Multipole Method of Greengard and Rokhlin, a general technique for efficient computation of energies and forces in systems of interacting particles. The basic idea is to decompose the particles into groups; particles within a group interact normally (i.e., by the full potential), whereas the energies and forces between groups of particles are calculated from their multipole moments. The efficiency of the fast multipole method is generally similar to that of Ewald summation, but is superior if the particles are clustered, i.e., if the system has large density fluctuations.

Multipole expansion of a potential outside an electrostatic charge distribution

Consider a discrete charge distribution consisting of N point charges q_i with position vectors $\mathbf{r}_i$. We assume the charges to be clustered around the origin, so that for all i: $r_i < r_{max}$, where r_{max} has some finite value. The potential $V(R)$, due to the charge distribution, at a point R outside the charge distribution, i.e., $|R| > r_{max}$, can be expanded in powers of $1/R$. Two ways of making this expansion can be found in the literature. The first is a Taylor series in the Cartesian coordinates x, y, and z, while the second is in terms of spherical harmonics which depend on spherical polar coordinates. The Cartesian approach has the advantage that no prior knowledge of Legendre functions, spherical harmonics, etc., is required. Its disadvantage is that the derivations are fairly cumbersome (in fact a large part of it is the implicit rederivation of the Legendre expansion of $1/|r\text{-}R|$, which was done once and for all by Legendre in the 1780s). Also it is difficult to give a closed expression for a general term of the multipole expansion—usually only the first few terms are given followed by an ellipsis.

Expansion in Cartesian Coordinates

The Taylor expansion of an arbitrary function $v(R\text{-}r)$ around the origin r = 0 is

$$v(\mathbf{R}-\mathbf{r}) = v(\mathbf{R}) - \sum_{\alpha=x,y,z} r_\alpha v_\alpha(\mathbf{R}) + \frac{1}{2}\sum_{\alpha=x,y,z}\sum_{\beta=x,y,z} r_\alpha r_\beta v_{\alpha\beta}(\mathbf{R}) - \cdots + \cdots$$

with

$$v_\alpha(\mathbf{R}) \equiv \left(\frac{\partial v(\mathbf{r}-\mathbf{R})}{\partial r_\alpha}\right)_{\mathbf{r}=\mathbf{0}} \quad \text{and} \quad v_{\alpha\beta}(\mathbf{R}) \equiv \left(\frac{\partial^2 v(\mathbf{r}-\mathbf{R})}{\partial r_\alpha \partial r_\beta}\right)_{\mathbf{r}=\mathbf{0}}.$$

If $v(r\text{-}R)$ satisfies the Laplace equation

$$\left(\nabla^2 v(\mathbf{r}-\mathbf{R})\right)_{\mathbf{r}=\mathbf{0}} = \sum_{\alpha=x,y,z} v_{\alpha\alpha}(\mathbf{R}) = 0$$

then the expansion can be rewritten in terms of the components of a traceless Cartesian second rank tensor:

$$\sum_{\alpha=x,y,z}\sum_{\beta=x,y,z} r_\alpha r_\beta v_{\alpha\beta}(\mathbf{R}) = \frac{1}{3}\sum_{\alpha=x,y,z}\sum_{\beta=x,y,z}(3r_\alpha r_\beta - \delta_{\alpha\beta} r^2) v_{\alpha\beta}(\mathbf{R}),$$

where $\delta_{\alpha\beta}$ is the Kronecker delta and r^2 ad $|r|^2$. Removing the trace is common, because it takes the rotationally invariant r^2 out of the second rank tensor.

Example

Consider now the following form of $v(r\text{-}R)$:

$$v(\mathrm{r}-\mathrm{R}) \equiv \frac{1}{|\mathrm{r}-\mathrm{R}|}.$$

Then by direct differentiation it follows that

$$v(\mathbf{R}) = \frac{1}{R}, \quad v_\alpha(\mathbf{R}) = -\frac{R_\alpha}{R^3}, \quad \text{and} \quad v_{\alpha\beta}(\mathbf{R}) = \frac{3R_\alpha R_\beta - \delta_{\alpha\beta}R^2}{R^5}.$$

Define a monopole, dipole, and (traceless) quadrupole by, respectively,

$$q_{\text{tot}} \equiv \sum_{i=1}^{N} q_i, \quad P_\alpha \equiv \sum_{i=1}^{N} q_i r_{i\alpha}, \quad \text{and} \quad Q_{\alpha\beta} \equiv \sum_{i=1}^{N} q_i (3r_{i\alpha} r_{i\beta} - \delta_{\alpha\beta} r_i^2),$$

and we obtain finally the first few terms of the multipole expansion of the total potential, which is the sum of the Coulomb potentials of the separate charges:

$$4\pi\varepsilon_0 V(\mathbf{R}) \equiv \sum_{i=1}^{N} q_i v(\mathbf{r}_i - \mathbf{R})$$

$$= \frac{q_{\text{tot}}}{R} + \frac{1}{R^3} \sum_{\alpha=x,y,z} P_\alpha R_\alpha + \frac{1}{6R^5} \sum_{\alpha,\beta=x,y,z} Q_{\alpha\beta}(3R_\alpha R_\beta - \delta_{\alpha\beta}R^2) + \cdots$$

This expansion of the potential of a discrete charge distribution is very similar to the one in real solid harmonics given below. The main difference is that the present one is in terms of linear dependent quantities, for

$$\sum_\alpha v_{\alpha\alpha} = 0 \quad \text{and} \quad \sum_\alpha Q_{\alpha\alpha} = 0.$$

NOTE: If the charge distribution consists of two charges of opposite sign which are an infinitesimal distance d apart, so that $d/R >> (d/R)^2$, it is easily shown that the only non-vanishing term in the expansion is

$$V(\mathbf{R}) = \frac{1}{4\pi\varepsilon_0 R^3}(\mathbf{P}\cdot\mathbf{R}),$$

the electric dipolar potential field.

Spherical Form

The potential V(R) at a point R outside the charge distribution, i.e., $|\mathrm{R}| > r_{\max}$, can be expanded by the Laplace expansion:

$$V(\mathbf{R}) \equiv \sum_{i=1}^{N} \frac{q_i}{4\pi\varepsilon_0 |\mathbf{r}_i - \mathbf{R}|} = \frac{1}{4\pi\varepsilon_0} \sum_{\ell=0}^{\infty} \sum_{m=-\ell}^{\ell} (-1)^m I_\ell^{-m}(\mathbf{R}) \sum_{i=1}^{N} q_i R_\ell^m(\mathbf{r}_i),$$

where $I_\ell^{-m}(\mathbf{R})$ is an irregular solid harmonic (defined below as a spherical harmonic function divided by $R^{\ell+1}$) and $R_\ell^m(\mathbf{r})$ is a regular solid harmonic (a spherical harmonic times r^ℓ). We define the *spherical multipole moment* of the charge distribution as follows

$$Q_\ell^m \equiv \sum_{i=1}^{N} q_i R_\ell^m(\mathbf{r}_i), \qquad -\ell \le m \le \ell.$$

Note that a multipole moment is solely determined by the charge distribution (the positions and magnitudes of the N charges).

A spherical harmonic depends on the unit vector $\hat{R}$. (A unit vector is determined by two spherical polar angles.) Thus, by definition, the irregular solid harmonics can be written as

$$I_\ell^m(\mathbf{R}) \equiv \sqrt{\frac{4\pi}{2\ell+1}} \frac{Y_\ell^m(\hat{R})}{R^{\ell+1}}$$

so that the *multipole expansion* of the field V(R) at the point R outside the charge distribution is given by

$$V(\mathbf{R}) = \frac{1}{4\pi\varepsilon_0} \sum_{\ell=0}^{\infty} \sum_{m=-\ell}^{\ell} (-1)^m I_\ell^{-m}(\mathbf{R}) Q_\ell^m$$

$$= \frac{1}{4\pi\varepsilon_0} \sum_{\ell=0}^{\infty} \left[\frac{4\pi}{2\ell+1}\right]^{1/2} \frac{1}{R^{\ell+1}} \sum_{m=-\ell}^{\ell} (-1)^m Y_\ell^{-m}(\hat{R}) Q_\ell^m, \qquad R > r_{\text{max}}.$$

This expansion is completely general in that it gives a closed form for all terms, not just for the first few. It shows that the spherical multipole moments appear as coefficients in the $1/R$ expansion of the potential.

It is of interest to consider the first few terms in real form, which are the only terms commonly found in undergraduate textbooks.

Since the summand of the m summation is invariant under a unitary transformation of both factors simultaneously and since transformation of complex spherical harmonics to real form is by a unitary transformation, we can simply substitute real irregular solid harmonics and real multipole moments. The $\ell = 0$ term becomes

$$V_{\ell=0}(\mathbf{R}) = \frac{q_{\text{tot}}}{4\pi\varepsilon_0 R} \qquad \text{with} \quad q_{\text{tot}} \equiv \sum_{i=1}^{N} q_i.$$

This is in fact Coulomb's law again. For the $\ell = 1$ term we introduce

$$\mathbf{R} = (R_x, R_y, R_z), \quad \mathbf{P} = (P_x, P_y, P_z) \quad \text{with} \quad P_\alpha \equiv \sum_{i=1}^{N} q_i r_{i\alpha}, \quad \alpha = x, y, z.$$

Then

$$V_{\ell=1}(\mathbf{R}) = \frac{1}{4\pi\varepsilon_0 R^3}(R_x P_x + R_y P_y + R_z P_z) = \frac{\mathbf{R}\cdot\mathbf{P}}{4\pi\varepsilon_0 R^3} = \frac{\hat{R}\cdot\mathbf{P}}{4\pi\varepsilon_0 R^2}.$$

This term is identical to the one found in Cartesian form.

In order to write the $\ell = 2$ term, we have to introduce shorthand notations for the five real components of the quadrupole moment and the real spherical harmonics.

Notations of the type

$$Q_{z^2} \equiv \sum_{i=1}^{N} q_i \frac{1}{2}(3z_i^2 - r_i^2),$$

can be found in the literature. Clearly the real notation becomes awkward very soon, exhibiting the usefulness of the complex notation.

Interaction of two non-overlapping charge distributions

Consider two sets of point charges, one set $\{q_i\}$ clustered around a point A and one set $\{q_j\}$ clustered around a point B.

Think for example of two molecules, and recall that a molecule by definition consists of electrons (negative point charges) and nuclei (positive point charges). The total electrostatic interaction energy U_{AB} between the two distributions is

$$U_{AB} = \sum_{i\in A}\sum_{j\in B}\frac{q_i q_j}{4\pi\varepsilon_0 r_{ij}}.$$

This energy can be expanded in a power series in the inverse distance of A and B. This expansion is known as the multipole expansion of U_{AB}.

In order to derive this multipole expansion, we write $r_{XY} = r_Y - r_X$, which is a vector pointing from X towards Y. Note that

$$\mathbf{R}_{AB} + \mathbf{r}_{Bj} + \mathbf{r}_{ji} + \mathbf{r}_{iA} = 0 \quad \Leftrightarrow \quad \mathbf{r}_{ij} = \mathbf{R}_{AB} - \mathbf{r}_{Ai} + \mathbf{r}_{Bj}.$$

We assume that the two distributions do not overlap:

$$|\,\mathbf{R}_{AB}\,| > |\,\mathbf{r}_{Bj} - \mathbf{r}_{Ai}\,| \quad \text{for all} \quad i, j.$$

Under this condition we may apply the Laplace expansion in the following form

$$\frac{1}{|\,\mathbf{r}_j - \mathbf{r}_i\,|} = \frac{1}{|\,\mathbf{R}_{AB} - (\mathbf{r}_{Ai} - \mathbf{r}_{Bj})\,|} = \sum_{L=0}^{\infty}\sum_{M=-L}^{L} (-1)^M I_L^{-M}(\mathbf{R}_{AB})\, R_L^M(\mathbf{r}_{Ai} - \mathbf{r}_{Bj}),$$

where I_L^M and R_L^M are irregular and regular solid harmonics, respectively. The translation of the regular solid harmonic gives a finite expansion,

$$R_L^M(\mathbf{r}_{Ai} - \mathbf{r}_{Bj}) = \sum_{\ell_A=0}^{L} (-1)^{L-\ell_A} \binom{2L}{2\ell_A}^{1/2}$$

$$\times \sum_{m_A=-\ell_A}^{\ell_A} R_{\ell_A}^{m_A}(\mathbf{r}_{Ai}) R_{L-\ell_A}^{M-m_A}(\mathbf{r}_{Bj}) \langle \ell_A, m_A; L-\ell_A, M-m_A \mid LM \rangle,$$

where the quantity between pointed brackets is a Clebsch-Gordan coefficient. Further we used

$$R_\ell^m(-\mathbf{r}) = (-1)^\ell R_\ell^m(\mathbf{r}).$$

Use of the definition of spherical multipoles $Q^m{}_l$ and covering of the summation ranges in a somewhat different order (which is only allowed for an infinite range of L) gives finally

$$U_{AB} = \frac{1}{4\pi\varepsilon_0} \sum_{\ell_A=0}^{\infty} \sum_{\ell_B=0}^{\infty} (-1)^{\ell_B} \binom{2\ell_A + 2\ell_B}{2\ell_A}^{1/2}$$

$$\times \sum_{m_A=-\ell_A}^{\ell_A} \sum_{m_B=-\ell_B}^{\ell_B} (-1)^{m_A+m_B} I_{\ell_A+\ell_B}^{-m_A-m_B}(\mathbf{R}_{AB})\, Q_{\ell_A}^{m_A} Q_{\ell_B}^{m_B} \langle \ell_A, m_A; \ell_B, m_B \mid \ell_A + \ell_B, m_A + m_B \rangle.$$

This is the multipole expansion of the interaction energy of two non-overlapping charge distributions which are a distance R_{AB} apart. Since

$$I_{\ell_A+\ell_B}^{-(m_A+m_B)}(\mathbf{R}_{AB}) \equiv \left[\frac{4\pi}{2\ell_A + 2\ell_B + 1}\right]^{1/2} \frac{Y_{\ell_A+\ell_B}^{-(m_A+m_B)}(\widehat{\mathbf{R}}_{AB})}{R_{AB}^{\ell_A+\ell_B+1}}$$

this expansion is manifestly in powers of $1/R_{AB}$. The function $Y^m{}_l$ is a normalized spherical harmonic.

Molecular Moments

All atoms and molecules (except S-state atoms) have one or more non-vanishing permanent multipole moments. Different definitions can be found in the literature, but the following definition in spherical form has the advantage that it is contained in one general equation. Because it is in complex form it has as the further advantage that it is easier to manipulate in calculations than its real counterpart.

We consider a molecule consisting of N particles (electrons and nuclei) with charges eZ_i. (Electrons have the Z-value unity, for nuclei it is the atomic number). Particle i has spherical polar coordinates r_i, θ_i, and φ_i and Cartesian coordinates x_i, y_i, and z_i. The (complex) electrostatic multipole operator is

$$Q_\ell^m \equiv \sum_{i=1}^{N} eZ_i \, R_\ell^m(\mathbf{r}_i),$$

where $R_\ell^m(\mathbf{r}_i)$ is a regular solid harmonic function in Racah's normalization (also known as Schmidt's semi-normalization). If the molecule has total normalized wave function ψ (depending on the coordinates of electrons and nuclei), then the multipole moment of order ℓ of the molecule is given by the expectation (expected) value:

$$M_\ell^m \equiv \langle \Psi \mid Q_\ell^m \mid \Psi \rangle.$$

If the molecule has certain point group symmetry, then this is reflected in the wave function: ψ transforms according to a certain irreducible representation λ of the group ("ψ has symmetry type λ"). This has the consequence that selection rules hold for the expectation value of the multipole operator, or in other words, that the expectation value may vanish because of symmetry. A well-known example of this is the fact that molecules with an inversion center do not carry a dipole (the expectation values of Q_1^m vanish for $m = -1, 0, 1$). For a molecule without symmetry no selection rules are operative and such a molecule will have non-vanishing multipoles of any order (it will carry a dipole and simultaneously a quadrupole, octupole, hexadecapole, etc.).

The lowest explicit forms of the regular solid harmonics (with the Condon-Shortley phase) give:

$$M_0^0 = \sum_{i=1}^{N} eZ_i,$$

(the total charge of the molecule). The (complex) dipole components are:

$$M_1^1 = -\sqrt{\tfrac{1}{2}} \sum_{i=1}^{N} eZ_i \langle \Psi \mid x_i + iy_i \mid \Psi \rangle \quad \text{and} \quad M_1^{-1} = \sqrt{\tfrac{1}{2}} \sum_{i=1}^{N} eZ_i \langle \Psi \mid x_i - iy_i \mid \Psi \rangle.$$

$$M_1^0 = \sum_{i=1}^{N} eZ_i \langle \Psi \mid z_i \mid \Psi \rangle.$$

Note that by a simple linear combination one can transform the complex multipole operators to real ones. The real multipole operators are of cosine type C_ℓ^m or sine type S_ℓ^m. A few of the lowest ones are:

$$C_1^0 = \sum_{i=1}^{N} eZ_i \, z_i$$

$$C_1^1 = \sum_{i=1}^{N} eZ_i \, x_i$$

$$S_1^1 = \sum_{i=1}^{N} eZ_i \, y_i$$

$$C_2^0 = \frac{1}{2}\sum_{i=1}^{N} eZ_i \, (3z_i^2 - r_i^2)$$

$$C_2^1 = \sqrt{3}\sum_{i=1}^{N} eZ_i \, z_i x_i$$

$$C_2^2 = \frac{1}{3}\sqrt{3}\sum_{i=1}^{N} eZ_i \, (x_i^2 - y_i^2)$$

$$S_2^1 = \sqrt{3}\sum_{i=1}^{N} eZ_i \, z_i y_i$$

$$S_2^2 = \frac{2}{3}\sqrt{3}\sum_{i=1}^{N} eZ_i \, x_i y_i$$

Note on Conventions

The definition of the complex molecular multipole moment given above is the complex conjugate of the definition given in this article, which follows the definition of the standard textbook on classical electrodynamics by Jackson, except for the normalization. Moreover, in the classical definition of Jackson the equivalent of the N-particle quantum mechanical expectation value is an integral over a one-particle charge distribution.

Remember that in the case of a one-particle quantum mechanical system the expectation value is nothing but an integral over the charge distribution (modulus of wavefunction squared), so that the definition of this article is a quantum mechanical N-particle generalization of Jackson's definition.

The definition in this article agrees with, among others, the one of Fano and Racah and Brink and Satchler.

Examples of Multipole Expansions

There are many types of multipole moments, since there are many types of potentials and many ways of approximating a potential by a series expansion, depending on the coordinates and the symmetry of the charge distribution. The most common expansions include:

- Axial multipole moments of a $1/R$ potential;
- Spherical multipole moments of a $1/R$ potential; and
- Cylindrical multipole moments of a ln R potential

Examples of $1/R$ potentials include the electric potential, the magnetic potential and the gravitational potential of point sources. An example of a ln R potential is the electric potential of an infinite line charge.

General Mathematical Properties

Multipole moments in mathematics and mathematical physics form an orthogonal basis for the decomposition of a function, based on the response of a field to point sources that are brought infinitely close to each other. These can be thought of as arranged in various geometrical shapes, or, in the sense of distribution theory, as directional derivatives.

Multipole expansions are related to the underlying rotational symmetry of the physical laws and their associated differential equations.

Even though the source terms (such as the masses, charges, or currents) may not be symmetrical, one can expand them in terms of irreducible representations of the rotational symmetry group, which leads to spherical harmonics and related sets of orthogonal functions. One uses the technique of separation of variables to extract the corresponding solutions for the radial dependencies.

In practice, many fields can be well approximated with a finite number of multipole moments (although an infinite number may be required to reconstruct a field exactly).

A typical application is to approximate the field of a localized charge distribution by its monopole and dipole terms. Problems solved once for a given order of multipole moment may be linearly combined to create a final approximate solution for a given source.

The Historical Development of Geo Magnetism

The history of geomagnetism is concerned with the history of the study of Earth's magnetic field. It encompasses the history of navigation using compasses, studies of the prehistoric magnetic field (archeomagnetism and paleomagnetism), and applications to plate tectonics.

Magnetism has been known since prehistory, but knowledge of the Earth's field developed slowly. The horizontal direction of the Earth's field was first measured in the fourth century BC but the vertical direction was not measured until 1544 AD and the intensity was first measured in 1791. At first, compasses were thought to point towards locations in the heavens, then towards magnetic mountains. A modern experimental approach to understanding the Earth's field began with *de Magnete*, a book published by William Gilbert in 1600. His experiments with a magnetic model of the Earth convinced him that the Earth itself is a large magnet.

Figure: *A reconstruction of an early Chinese compass. A spoon made of lodestone, its handle pointing south, was mounted on a brass plate with astrological symbols.*

Early Ideas on Magnetism

Knowledge of the existence of magnetism probably dates back to the prehistoric development of iron smelting. Iron can be obtained on the Earth's surface from meteorites; the mineral lodestone is rich in the magnetic mineral magnetite and can be magnetized by a lightning strike. In his *Natural History*, Pliny the Elder recounts a legend about a shepherd called Magnes on the island of Crete whose iron-studded boots kept sticking to the path. The earliest ideas on the nature of magnetism are attributed to Thales (c. 624 BC – c. 546 BC).

In classical antiquity, little was known about the nature of magnetism. No sources mention the two poles of a magnet or its tendency to point northward. There were two main theories about the origins of magnetism. One, proposed by Empedocles of Acragas and taken up by Plato and Plutarch, invoked an invisible *effluvium* seeping through the pores of materials; Democritus of Abdera replaced this effluvium by atoms, but the mechanism was essentially the same. The other theory evoked the metaphysical principle of *sympathy* between similar objects. This was mediated by a purposeful life force that strove toward perfection. This theory can be found in the writings of Pliny the Elder and Aristotle, who claimed that Thales attributed a soul to the magnet. In China, a similar life force, or *qi*, was believed to animate magnets, so the Chinese used early compasses for feng shui.

Little changed in the view of magnetism during the Middle Ages, and some classical ideas lingered until well after the first scientific experiments on magnetism. One belief, dating back to Pliny, was that fumes from eating garlic and onions could destroy the magnetism in a compass, rendering it useless. Even after William Gilbert disproved this in 1600, there were reports of helmsmen on British ships being flogged for eating garlic. However, this belief was far from universal. In 1558 Giambattista della Porta reported "When I enquired of mariners whether it were so that they were forbid to eat onyones and garlick for that reason, they said they were old wives fables and things ridiculous, and that sea-men would sooner lose their lives then abstain from eating onyons and garlick."

Measurement of the Field

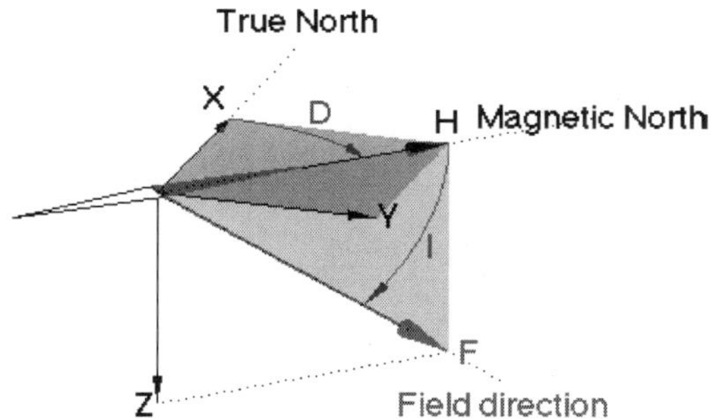

Figure: *Illustration of the coordinate systems used for representing the Earth's magnetic field. The coordinates X,Y,Z correspond to North, East and down; D is the declination and I is the inclination.*

At a given location, a full representation of the Earth's magnetic field requires a vector with three coordinates. These can be Cartesian (North, East and Down) or spherical (declination, inclination and intensity). In the latter system, the declination (the deviation from true north, a horizontal angle) must be measured first to establish the direction of magnetic North; then the dip (a vertical angle) can be measured relative to magnetic North. In China, the horizontal direction was measured as early as the fourth century BC, and the existence of declination first recognised in 1088.

In Europe, this was not widely accepted until the middle of the fifteenth century AD. Inclination (also known as *magnetic dip*) was first measured in 1544 AD. The intensity was not measured until 1791, after advances in the understanding of electromagnetism.

Compass

Figure: *A simple dry magnetic portable compass.*

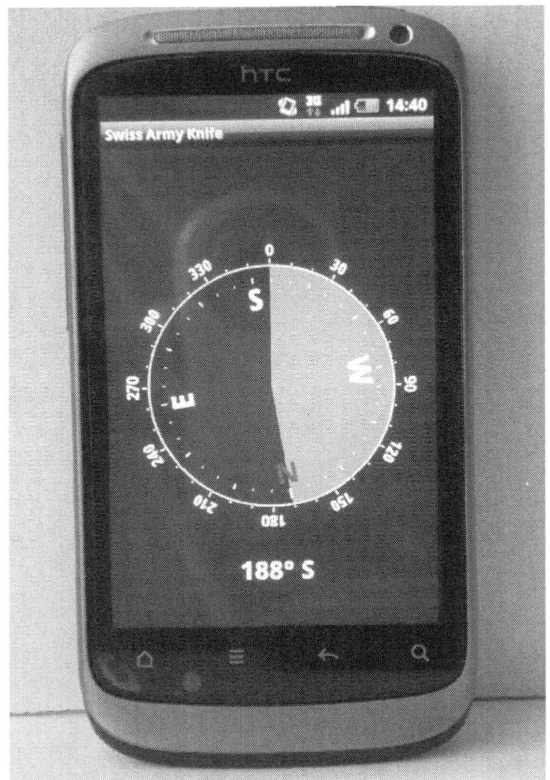

Figure: *A smartphone that can be used as a compass because of the magnetometer inside.*

A compass is a navigational instrument that shows directions in a frame of reference that is stationary relative to the surface of the Earth. The frame of reference defines the four *cardinal directions* (or *points*) – north, south, east, and west. Intermediate directions are also defined. Usually, a diagram called a compass rose, which shows the directions (with their names usually abbreviated to initials), is marked on the compass. When the compass is in use, the rose is aligned with the real directions in the frame of reference, so, for example, the "N" mark on the rose really points to the north.

Frequently, in addition to the rose or sometimes instead of it, angle markings in degrees are shown on the compass. North corresponds to zero degrees, and the angles increase clockwise, so east is 90 degrees, south is 180, and west is 270. These numbers allow the compass to show azimuths or bearings, which are commonly stated in this notation.

The magnetic compass was first invented as a device for divination as early as the Chinese Han Dynasty (since about 206 BC). The compass was used in Song Dynasty China by the military for navigational orienteering by 1040-1044, and was used for maritime navigation by 1111 to 1117. The use of a compass is recorded in Western Europe between 1187 and 1202, and in Persia in 1232. The dry compass was invented in Europe around 1300. This was supplanted in the early 20th century by the liquid-filled magnetic compass.

Types of Compasses

There are two widely used and radically different types of compass. The magnetic compass contains a magnet that interacts with the earth's magnetic field and aligns itself to point to the magnetic poles. Simple compasses of this type show directions in a frame of reference in which the directions of the magnetic poles are due north and south. These directions are called *magnetic north* and *magnetic south*. The gyro compass (sometimes spelled with a hyphen, or as one word) contains a rapidly spinning wheel whose rotation interacts dynamically with the rotation of the earth so as to make the wheel process, losing energy to friction until its axis of rotation is parallel with the earth's.

The wheel's axis therefore points to the earth's rotational poles, and a frame of reference is used in which the directions of the rotational poles are due north and south. These directions are called *true north* and *true south*, respectively. The astrocompass works by observing the direction of stars and other celestial bodies.

There are other devices which are not conventionally called compasses but which do allow the true cardinal directions to be determined. Some GPS receivers have two or three antennas, fixed some distance apart to the structure of a vehicle, usually an aircraft or ship. The exact latitudes and longitudes of the antennas can be determined simultaneously, which allows the directions of the cardinal points to be calculated relative to the heading of the aircraft (the direction in which its nose is pointing), rather than to its direction of movement, which will be different if there is a crosswind. They are said to work "like a compass", or "as a compass".

Figure: *A military compass that was used during World War I.*

Even a GPS device or similar can be used as compass, since if the receiver is being moved, even at walking pace, it can follow the change of its position, and hence determine the compass bearing of its direction of movement, and hence the directions of the cardinal points relative to its direction of movement. A much older example was the Chinese south-pointing chariot, which worked like a compass by directional dead reckoning. It was initialized by hand, possibly using astronomical observations e.g. of the Pole Star, and thenceforth counteracted every turn that was made to keep its pointer aiming in the desired direction, usually to the south. Watches and sundials can also be used to find compass directions.

A recent development is the electronic compass which detects the direction without potentially fallible moving parts. This may use a fibre optic gyrocompass or a magnetometer. The magnetometer frequently appears as an optional subsystem built into hand-held GPS receivers and mobile phones. However, magnetic compasses remain popular, especially in remote areas, as they are relatively inexpensive, durable, and require no power supply.

Magnetic Compass

The magnetic compass consists of a magnetized pointer (usually marked on the North end) free to align itself with Earth's magnetic field. A compass is any magnetically sensitive device capable of indicating the direction of the magnetic north of a planet's magnetosphere. The face of the compass generally highlights the cardinal points of north, south, east and west. Often, compasses are built as a stand alone sealed instrument with a magnetized bar or needle turning freely upon a pivot, or moving in a fluid, thus able to point in a northerly and southerly direction.

The compass greatly improved the safety and efficiency of travel, especially ocean travel. A compass can be used to calculate heading, used with a sextant to calculate latitude, and with a marine chronometer to calculate longitude. It thus provides a much improved navigational capability that has only been recently supplanted by modern devices such as the Global Positioning System (GPS).

How a Magnetic Compass Works

A compass functions as a pointer to "magnetic north" because the magnetized needle at its heart aligns itself with the lines of the Earth's magnetic field. The magnetic field exerts a torque on the needle, pulling one end or *pole* of the needle toward the Earth's North magnetic pole, and the other toward the South magnetic pole. The needle is mounted on a low-friction pivot point, in better compasses a jewel bearing, so it can turn easily. When the compass is held level, the needle turns until, after a few seconds to allow oscillations to die out, one end points toward the North magnetic pole.

A magnet or compass needle's "north" pole is defined as the one which is attracted to the North magnetic pole of the Earth. Since opposite poles attract ("north" to "south") the North magnetic pole of the Earth is actually the *south* pole of the Earth's magnetic field. The compass needle's north pole is always marked in some way: with a distinctive colour, luminous paint, or an arrowhead.

Figure: *An inexpensive compass, aligned so that its needle points through the "North" mark on its compass card.*

Instead of a needle, professional compasses usually have bar magnets glued to the underside of a disk pivoted in the center so it can turn, called a "compass card", with a "compass rose" showing the cardinal points and degrees marked on it. Better compasses are "*liquid-filled*"; the chamber containing the needle or disk is filled with a liquid whose purpose is to damp the oscillations of the needle so it will settle down to point to North more quickly, and also to protect the needle or disk from shock.

In navigation, directions on maps are expressed with reference to *geographical* or *true north*, the direction toward the Geographical North Pole, the rotation axis of the Earth. Since the Earth's magnetic poles are near, but are not at the same locations as its geographic poles, a compass does not point to true north.

The direction a compass points is called *magnetic north*, the direction of the North magnetic pole. Depending on where the compass is located on the surface of the Earth the angle between true north and magnetic north, called *magnetic declination* can vary widely, increasing the farther one is from the prime meridian of the Earth's magnetic field. The local magnetic declination is given on most maps, to allow the map to be oriented with a compass parallel to true north. Some magnetic compasses include means to manually compensate for the magnetic declination, so that the compass shows true directions.

In geographic regions near the magnetic poles, in the Arctic and Antarctic, variations in the Earth's magnetic field cause magnetic compasses to have such large errors that they are useless, so other instruments must be used for navigation.

The positions of the magnetic poles change over time, on a time-scale that is not extremely long by human standards. They wander over time-periods of only a few years, leading to corresponding changes of the directions of magnetic north and south as observed everywhere on the planet.

History

The compass was invented in China, during the Han Dynasty between the 2nd century BC and 1st century AD. The first compasses were made of lodestone, a naturally magnetized ore of iron. Ancient Chinese people found that if a lodestone was suspended so it could turn freely, it would always point in the same direction, toward the magnetic poles.

Early compasses were used for geomancy "in the search for gems and the selection of sites for houses," but were later adapted for navigation during the Song Dynasty in the 11th century. Later compasses were made of iron needles, magnetized by striking them with a lodestone.

The dry compass was invented in medieval Europe around 1300. This was supplanted in the early 20th century by the liquid-filled magnetic compass.

Navigation Prior to the Compass

Prior to the introduction of the compass, *position, destination,* and direction at sea were primarily determined by the sighting of landmarks, supplemented with the observation of the position of celestial bodies. On cloudy days, the Vikings may have used cordierite or some other birefringent crystal to determine the sun's direction and elevation from the polarization of daylight; their astronomical knowledge was sufficient to let them use this information to determine their proper heading. For more southerly Europeans unacquainted with this technique, the invention of the compass enabled the determination of heading when the sky was overcast or foggy. This enabled mariners to navigate safely far from land, increasing sea trade, and contributing to the Age of Discovery.

Geomancy and Feng Shui

Magnetism was originally used not for navigation, but for geomancy and fortune-telling by the Chinese. The earliest Chinese magnetic compasses were probably not designed for navigation, but rather to order and harmonize their environments and buildings in accordance with the geomantic principles of *feng shui*. These early compasses were made with lodestone, a form of the mineral magnetite that is a naturally-occurring magnet and aligns itself with the Earth's magnetic field.

Based on Krotser and Coe's discovery of an Olmec hematite artifact in Mesoamerica, radiocarbon dated to 1400-1000 BC, astronomer John Carlson has hypothesized that the Olmec might have used the geomagnetic lodestone earlier than 1000 BC for geomancy, a method of divination, which if proven true, predates the Chinese use of magnetism for feng shui by a millennium. Carlson speculates that the Olmecs used similar artifacts as a directional device for astronomical or geomantic purposes but does not suggest navigational usage.

The artifact is part of a polished hematite (lodestone) bar with a groove at one end (possibly for sighting). The artifact now consistently points 35.5 degrees west of north, but may have pointed north-south when whole. Carlson's claims have been disputed by other scientific researchers, who have suggested that the artifact is actually a constituent piece of a decorative ornament and not a purposely built compass. Several other hematite or magnetite artifacts have been found at pre-Columbian archaeological sites in Mexico and Guatemala.

Navigational Compass

A number of ancient cultures used lodestones, suspended so they could turn, as magnetic compasses for navigation. Early mechanical compasses are referenced in written records of the Chinese, who began using it for navigation sometime between the 9th and 11th century, "some time before 1050, possibly as early as 850." A common theory by historians, suggests that the Arabs introduced the compass from China to Europe, although current textual evidence only supports the fact that Chinese use of the navigational compass preceded that of Europe and the Middle East.

Declination

The magnetic compass existed in China back as far as the fourth century BC. It was used as much for feng shui as for navigation on land. It was not until good steel needles could be forged that compasses were used for navigation at sea; before that, they could not retain their magnetism for long. The existence of magnetic declination, the difference between magnetic north and true north, was first recognised by Shen Kuo in 1088.

The first mention of a compass in Europe was in 1190 AD by Alexander Neckham. He described it as a common navigational aid for sailors, so the compass must have been introduced to Europe some time earlier. Whether the knowledge came from China to Europe, or was invented separately, is not clear. If the knowledge was transmitted, the most likely intermediary was Arab merchants, but Arabic literature does not mention the compass until after Neckham. There is also a difference in convention: Chinese compasses point south while European compasses point north.

In 1269, Pierre de Maricourt (commonly referred to as *Petrus Peregrinus*) wrote a letter to a friend in which he described two kinds of compass, one in which an oval lodestone floated in a bowl of water, and the first dry compass with the needle mounted on a pivot. He also was the first to write about experiments with magnetism and describe the laws of attraction. An example is the experiment where a magnet is broken into two pieces and the two pieces can attract and repel each other (in modern terms, they both have north and south poles). This letter, generally referred to as *Epistola de Magnete*, was a landmark in the history of science.

Petrus Peregrinus assumed that compasses point towards true north. While his contemporary Roger Bacon is reputed to observe that compasses deviated from true north, the idea of magnetic declination was only gradually accepted. At first it was thought that the declination must be the result of systematic error. However, by the middle of the fifteenth century, sundials in Germany were oriented using corrections for declination.

Inclination

A compass must be balanced to counter the tendency of the needle to dip in the direction of the Earth's field. Otherwise, it will not spin freely. Often, compasses that are balanced for one latitude do not work as well at a different latitude. This problem was first reported by Georg Hartmann, a vicar in Nuremberg, in 1544. Robert Norman was the first to recognise that this occurs because the Earth's field itself is tilted from the vertical. In his book *The Newe Attractive*, Norman called inclination "a newe discovered secret and subtle property concerning the Declining of the Needle." He created a compass in which the needle was floated in a goblet of water, attached to a cork to make it neutrally buoyant. The needle could orient itself in any direction, so it dipped to align itself with the Earth's field. Norman also created a dip circle, a compass needle pivoted about a horizontal axis, to measure the effect.

Early Ideas about the Source

In early attempts to understand the Earth's magnetic field, measuring it was only part of the challenge. Understanding the measurements was also difficult because the mathematical and physical concepts had not yet been developed – in particular, the concept of a vector field that associates a vector with each point in space. The Earth's field is generally represented by field lines that run from pole to pole; the field at any point is parallel to a field line but does not have to point at either pole.

As late as the eighteenth century, however, a natural philosopher would believe that a magnet had to be pointing directly at something. Thus, the Earth's magnetic field had to be explained by localized sources, and as more was learned about the Earth's field, these sources became increasingly complex.

At first, in both China and Europe, the source was assumed to be in the heavens – either the celestial poles or the Pole star. These theories required that magnets point at (or very close to) true north, so they ran into difficulty when the existence of declination was accepted. Then natural philosophers began to propose earthly sources such as a rock or mountain.

Legends about magnetic mountains go back to the classical era. Ptolemy recounted a legend about magnetic islands (now thought to be near Borneo) that exerted such a strong attraction on ships with nails that the ships were held in place and could not move. Even more dramatic was the Arab legend (recounted in *One Thousand and One Nights*) that a magnetic mountain could pull all the nails out of a ship, causing the ship to fall apart and founder. The story passed to Europe and became part of several epic tales.

Europeans started to place magnetic mountains on their maps in the sixteenth century. A notable example is Gerardus Mercator, whose famous maps included a magnetic mountain or two near the North Pole. At first, he just placed a mountain in an arbitrary location; but later he attempted to measure its location based on declinations from different locations in Europe. When subsequent measurements resulted in two contradictory estimates for the mountain, he simply placed two mountains on the map.

Beginnings of Modern Science

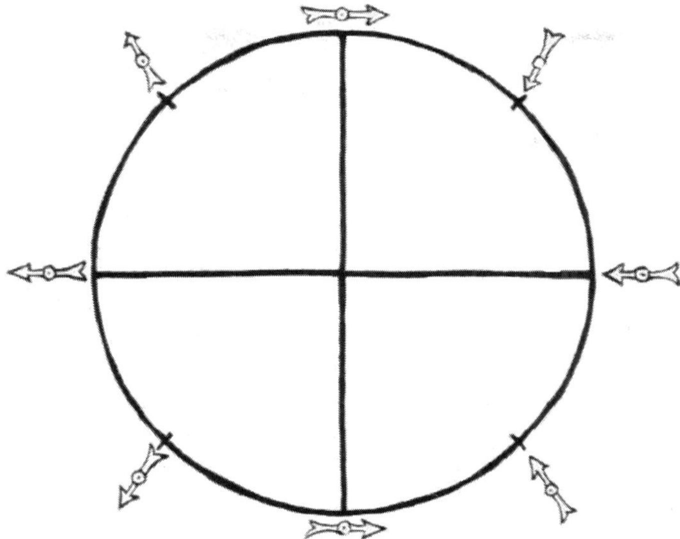

Figure: *An illustration of compass directions at various latitudes on the Earth from de Magnete. North is to the right.*

William Gilbert

Magnus magnes ipse est globus terrestris. (The Earth itself is a great magnet.)

—William Gilbert, *De Magnete*

1600 was a notable year for William Gilbert. He became president of the Royal College of Physicians of London, was appointed personal physician for Queen Elizabeth I, and wrote *De Magnete*, one of the books that mark the beginning of modern science. *De Magnete* is most famous for introducing (or at least popularizing) an experimental approach to science and deducing that the Earth is a great magnet.

Gilbert's book is divided into six chapters. The first is an introduction in which he discusses the importance of experiment and various facts about the Earth, including the insignificance of surface topography compared to the radius of the Earth. He also announces his deduction that the Earth is a great magnet. In book 2, Gilbert deals with "coition", or the laws of attraction. Gilbert distinguishes between magnetism and static electricity (the latter being induced by rubbing amber) and reports many experiments with both (some dating back to Peregrinus). One involves breaking a magnet in two and showing that both parts have a north and south pole. He also dismisses the idea of perpetual motion. The third book has a general description of magnetic directions along with details on how to magnetize a needle. He also introduces his *terella*, or "little Earth". This is a magnetized sphere that he uses to model the magnetic properties of the Earth. In chapters 4 and 5 he goes into more detail about the two components of the direction, declination and inclination.

In the late 1590s Henry Briggs, a professor of geometry at Gresham College in London, had published a table of magnetic inclination with latitude for the earth. It agreed well with the inclinations that Gilbert measured around the circumference of his terella. Gilbert deduced that the Earth's magnetic field is equivalent to that of a uniformly magnetized sphere, magnetized parallel to the axis of rotation (in modern terms, a *geocentric axial dipole*).

However, he was aware that declinations were not consistent with this model. Based on the declinations that were known at the time, he proposed that the continents, because of their raised topography, formed centers of attraction that made compass needles deviate. He even demonstrated this effect by gouging out some topography on his terella and measuring the effect on declinations.

A Jesuit monk, Niccolò Cabeo, later took a leaf from Gilbert's book and showed that, if the topography was on the correct scale for the Earth, the differences between the highs and lows would only be about one tenth of a millimeter. Therefore, the continents could not noticeably affect the declination.

The sixth book of *de Magnete* was devoted to cosmology. He dismissed the prevailing Ptolemaic model of the universe, in which the planets and stars are organised in a series of concentric shells rotating about the Earth, on the grounds that the speeds involved would be absurdly large ("there cannot be diurnal motion of infinity"). Instead, the Earth was rotating about its own axis.

In place of the concentric shells, he proposed that the heavenly bodies interacted with each other and Earth through magnetic forces. Magnetism maintained the Earths position and made it rotate, while the magnetic attraction of the Moon drove the tides.

Some obscure reasoning led to the peculiar conclusion that a terella, if freely suspended, would orient itself in the same direction as the Earth and rotate daily. Both Kepler and Galileo would adopt Gilbert's idea of magnetic attraction between heavenly bodies, but Newton's law of universal gravitation would render it obsolete.

Guillaume le Nautonier

In about 1603, the Frenchman Guillaume le Nautonier (William the Navigator), Sieur de Castelfranc, published a rival theory of the Earth's field in his book *Mecometrie de l'eymant (Measurement of longitude with a magnet).*

Le Nautonier was a mathematician, astronomer and Royal Geographer in the court of Henry IV. He disagreed with Gilbert's assumption that the Earth had to be magnetized parallel to the rotational axis, and instead produced a model in which the magnetic moment was tilted by 22.5° – in effect, the first tilted dipole model. The last 196 pages of his book were taken up with tables of latitudes and longitudes with declination and inclination for use by mariners. If his model had been accurate, it could have been used to determine both latitude and longitude using a combination of magnetic declination and astronomical observations.

Le Nautonier tried to sell his model to Henry IV, and his son to the English leader Oliver Cromwell, both without success. It was widely criticized, with Didier Dounot concluding that the work was based on "unfounded assumptions, errors in calculation and data manipulation". However, the geophysicist Jean-Paul Poirier examined the works of both le Nautonier and Dounot, and found that the error was in Dounot's reasoning.

Temporal Variation

One of Gilbert's conclusions was that the Earth's field could not vary in time. This was soon to be proved false by a series of measurements in London.

In 1580, William Borough measured the declination and found it to be $11/_4$° NE. In 1622, Edmund Gunter found it to be 5° 56' NE. He noted the difference from Borough's result but concluded that Borough must have made a measurement error.

In 1633, Henry Gellibrand measured the declination in the same location and found it to be 4° 05' NE. Because of the care with which Gunther had made his measurements, Gellibrand was confident that the changes were real. In 1635 he published *A Discourse Mathematical on the Variation of the Magneticall Needle* stating that the declination had changed by more than 7° in 54 years. The reality of geomagnetic secular variation was rapidly accepted in England, where Gellibrand had a high reputation, but in other countries it was met with skepticism until it was confirmed by further measurements.

Magnetic Navigation

Early mariners used portolan charts for navigation. These charts showed coastline with rhumb lines connecting ports. A mariner could navigate by aligning the chart with a compass and following the compass heading. Early charts had distorted coastlines because the cartographers did not know about declination, but the charts still worked because mariners were sailing in straight lines.

While boats mainly plied seas the size of the Mediterranean, rhumb lines were sufficient for navigation. However, when they ventured into the Atlantic and Pacific oceans, it was no longer sufficient to plot a straight-line course from one destination to another (although Christopher Columbus used dead reckoning and a fixed compass direction). Mariners needed to determine their latitude and longitude.

In the Age of Sail, dating from the sixteenth to the mid-nineteenth century, international trade was dominated by sailing ships. More than one European government offered a generous prize to the first person who could accurately determine longitude. The British prize, the longitude prize, led to the development of the marine chronometer.

Chapter 2

Magnetic Field that Extends into Space

A magnetosphere is the area of space near an astronomical object in which charged particles are controlled by that object's magnetic field. Near the surface of the object, the magnetic field lines resemble those of a magnetic dipole. Farther away from the surface, the field lines are significantly distorted by electric currents flowing in the plasma (e.g. in ionosphere or solar wind). When speaking about Earth, *magnetosphere* is typically used to refer to the outer layer of the ionosphere, although some sources consider the ionosphere and magnetosphere to be separate.

History

Study of Earth's magnetosphere began in 1600, when William Gilbert discovered that the magnetic field on the surface of Earth resembled that on a terrella, a small, magnetized sphere. In the 1940s, Walter M. Elsasser proposed the model of dynamo theory, which attributes Earth's magnetic field to the motion of Earth's iron outer core. Through the use of magnetometers, scientists were able to study the variations in Earth's magnetic field as functions of both time and latitude and longitude. Beginning in the late 1940s, rockets were used to study cosmic rays. In 1958, Explorer 1, the first of the Explorer series of space missions, was launched to study the intensity of cosmic rays above the atmosphere and measure the fluctuations in this activity. This mission observed the existence of the Van Allen radiation belt (located in the inner region of Earth's magnetosphere), with the Explorer 3 mission later that year definitively proving its existence.

Also in 1958, Eugene Parker proposed the idea of the solar wind. In 1959, the term magnetosphere was proposed by Thomas Gold.

The Explorer 12 mission in 1961 led to the observation by Cahill and Amazeen in 1963 of a sudden decrease in the strength of the magnetic field near the noon meridian, later named the magnetopause. In 1983, the International Cometary Explorer observed the magnetotail, or the distant magnetic field.

Types of Magnetospheres

The structure and behaviour of magnetospheres is dependent on several variables: the type of astronomical object, the nature of sources of plasma and momentum, the period of the object's spin, the nature of the axis where about the object spins, the axis of the magnetic dipole, and the magnitude and direction of the velocity of the flow of solar wind.

The distance at which a planet can withstand the solar wind pressure is called the Chapman–Ferraro distance. This is modelled by a formula wherein R_P represents the radius of the planet, B_{surf} represents the magnetic field on the surface of the planet at the equator, and V_{SW} represents the velocity of the solar wind.

$$R_{CF} = R_P \left(\frac{B_{surf}^2}{\mu_0 \rho V_{SW}^2} \right)^{\frac{1}{6}}$$

A magnetosphere is classified as "intrinsic" when $R_{CF} \gg R_P$, or when the primary opposition to the flow of solar wind is the magnetic field of the object. Mercury, Earth, Jupiter, Ganymede, Saturn, Uranus, and Neptune exhibit intrinsic magnetospheres. A magnetosphere is classified as "induced" when $R_{CF} \ll R_P$, or when the solar wind is not opposed by the object's magnetic field.

In this case, the solar wind interacts with the atmosphere or ionosphere of the planet (or surface of the planet, if the planet has no atmosphere).

Venus has an induced magnetic field. What this means is that because Venus appears to have no internal dynamo effect, the only magnetic field present is that formed by the solar wind's wrapping around the physical obstacle of Venus. When $R_{CF} \approx R_P$, the planet itself and its magnetic field both contribute. It is possible that Mars is of this type.

Structure

Bow Shock

The bow shock forms the outermost layer of the magnetosphere: the boundary between the magnetosphere and the ambient medium. For stars, this is usually the boundary between the stellar wind and interstellar medium; for planets, the speed of the solar wind there plummets as it approaches the magnetopause.

Magnetosheath

The magnetosheath is the region of the magnetosphere between the bow shock and the magnetopause. It is formed mainly from shocked solar wind, though it contains a small amount of plasma from the magnetosphere. It is an area exhibiting high particle energy flux, where the direction and magnitude of the magnetic field varies erratically. This is caused by the collection of solar wind gas that has effectively undergone thermalization. It acts as a cushion that transmits the pressure from the flow of the solar wind and the barrier of the magnetic field from the object.

Magnetopause

The magnetopause is the area of the magnetosphere wherein the pressure from the planetary magnetic field is balanced with the pressure from the solar wind. It is the convergence of the shocked solar wind from the magnetosheath with the magnetic field of the object and plasma from the magnetosphere.

Because both sides of this convergence contain magnetized plasma, the interactions between them are very complex. The structure of the magnetopause depends upon the Mach number and beta of the plasma, as well as the magnetic field. The magnetopause changes size and shape as the pressure from the solar wind fluctuates.

Magnetotail

Opposite the compressed magnetic field is the magnetotail, where the magnetosphere extends far beyond the astronomical object. It contains two lobes, referred to as the northern and southern tail lobes. The northern tail lobe points towards the object and the southern tail lobe points away.

The tail lobes are almost empty, with very few charged particles opposing the flow of the solar wind. The two lobes are separated by a plasma sheet, an area where the magnetic field is weaker and the density of charged particles is higher.

Earth's Magnetosphere

Over Earth's equator, the magnetic field lines become almost horizontal, then return to connect back again at high latitudes. However, at high altitudes, the magnetic field is significantly distorted by the solar wind and its solar magnetic field. On the dayside of Earth, the magnetic field is significantly compressed by the solar wind to a distance of approximately 65,000 kilometers (40,000 mi). Earth's bow shock is about 17 kilometers (11 mi) thick and located about 90,000 kilometers (56,000 mi) from Earth. The magnetopause exists at a distance of several hundred kilometers off earth's surface.

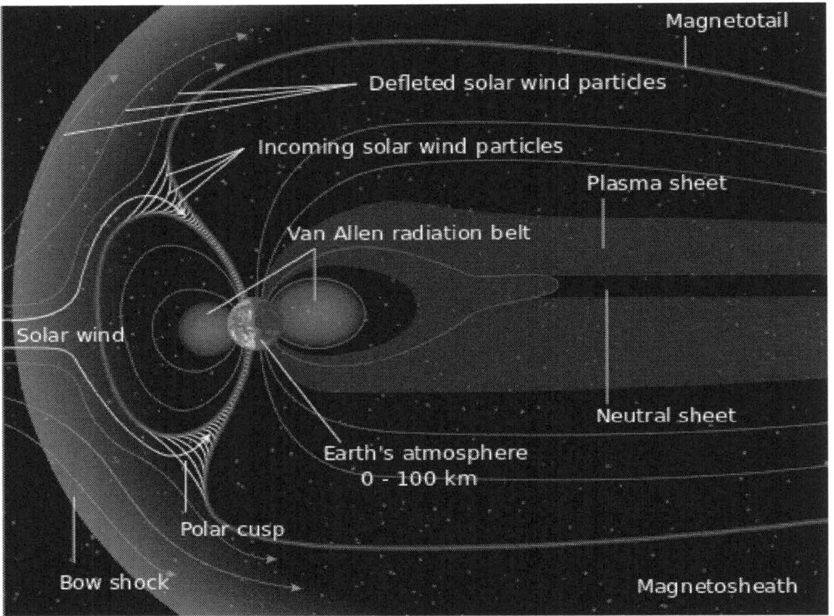

Earth's magnetopause has been compared to a sieve because it allows solar wind particles to enter. Kelvin–Helmholtz instabilities occur when large swirls of plasma travel along the edge of the magnetosphere at a different velocity from the magnetosphere, causing the plasma to slip past. This results in magnetic reconnection, and as the magnetic field lines break and reconnect, solar wind particles are able to enter the magnetosphere. On Earth's nightside, the magnetic field extends in the magnetotail, which lengthwise exceeds 6,300,000 kilometers (3,900,000 mi). Earth's magnetotail is the primary source of the polar aurora. Also, NASA scientists have suggested or "speculated" that Earth's magnetotail can cause "dust storms" on the Moon by creating a potential difference between the day side and the night side.

Other Objects

The magnetosphere of Jupiter is the largest planetary magnetosphere in the Solar System, extending up to 7,000,000 kilometers (4,300,000 mi) on the dayside and almost to the orbit of Saturn on the nightside. Jupiter's magnetosphere is stronger than Earth's by an order of magnitude, and its magnetic moment is approximately 18,000 times larger.

Stellar Magnetic Field

A stellar magnetic field is a magnetic field generated by the motion of conductive plasma inside a star. This motion is created through convection, which is a form of energy transport involving the physical movement of material. A localized magnetic field exerts a force on the plasma, effectively increasing the pressure without a comparable gain in density. As a result, the magnetized region rises relative to the remainder of the plasma, until it reaches the star's photosphere. This creates starspots on the surface, and the related phenomenon of coronal loops.

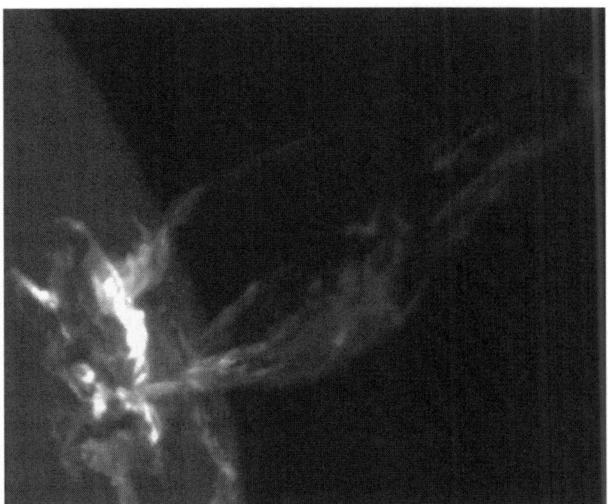

Figure: *The magnetic field of the Sun is driving this massive ejection of plasma. NOAA image.*

Measurement

The magnetic field of a star can be measured by means of the Zeeman effect. Normally the atoms in a star's atmosphere will absorb certain frequencies of energy in the electromagnetic spectrum, producing characteristic dark absorption lines in the spectrum. When the atoms are within a magnetic field, however, these lines become split into multiple, closely spaced lines. The energy also becomes polarized with an orientation that depends on orientation of the magnetic field. Thus the strength and direction of the star's magnetic field can be determined by examination of the Zeeman effect lines.

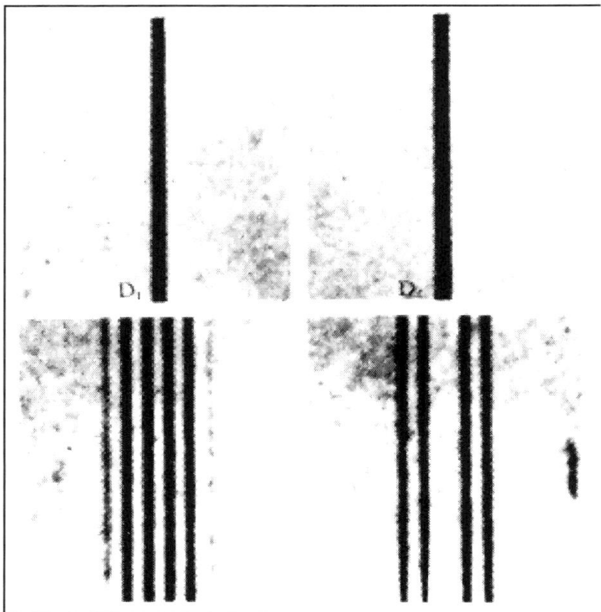

Figure: *The lower spectrum demonstrates the Zeeman effect after a magnetic field is applied to the source at top.*

A stellar spectropolarimeter is used to measure the magnetic field of a star. This instrument consists of a spectrograph combined with a polarimeter. The first instrument to be dedicated to the study of stellar magnetic fields was NARVAL, which was mounted on the Bernard Lyot Telescope at the Pic du Midi de Bigorre in the French Pyrenees mountains.

Various measurements—including magnetometer measurements over the last 150 years; ^{14}C in tree rings; and ^{10}Be in ice cores—have established substantial magnetic variability of the Sun on decadal, centennial and millennial time scales.

Field Generation

Stellar magnetic fields, according to solar dynamo theory, are caused within the convective zone of the star. The convective circulation of the conducting plasma functions like a dynamo. This activity destroys the star's primordial magnetic field, then generates a dipolar magnetic field. As the star undergoes differential rotation—rotating at different rates for various latitudes—the magnetism is wound into a toroidal field of "flux ropes" that become wrapped around the star. The fields can become highly concentrated, producing activity when they emerge on the surface.

The magnetic field of a rotating body of conductive gas or liquid develops self-amplifying electric currents, and thus a self-generated magnetic field, due to a combination of differential rotation (different angular velocity of different parts of body), Coriolis forces and induction. The distribution of currents can be quite complicated, with numerous open and closed loops, and thus the magnetic field of these currents in their immediate vicinity is also quite twisted. At large distances, however, the magnetic fields of currents flowing in opposite directions cancel out and only a net dipole field survives, slowly diminishing with distance. Because the major currents flow in the direction of conductive mass motion (equatorial currents), the major component of the generated magnetic field is the dipole field of the equatorial current loop, thus producing magnetic poles near the geographic poles of a rotating body.

The magnetic fields of all celestial bodies are often aligned with the direction of rotation, with notable exceptions such as certain pulsars. Another feature of this dynamo model is that the currents are AC rather than DC. Their direction, and thus the direction of the magnetic field they generate, alternates more or less periodically, changing amplitude and reversing direction, although still more or less aligned with the axis of rotation.

The Sun's major component of magnetic field reverses direction every 11 years (so the period is about 22 years), resulting in a diminished magnitude of magnetic field near reversal time. During this dormancy, the sunspots activity is at maximum (because of the lack of magnetic braking on plasma) and, as a result, massive ejection of high energy plasma into the solar corona and interplanetary space takes place. Collisions of neighbouring sunspots with oppositely directed magnetic fields result in the generation of strong electric fields near rapidly disappearing magnetic field regions. This electric field accelerates electrons and protons to high energies (kiloelectronvolts) which results in jets of extremely hot plasma leaving the Sun's surface and heating coronal plasma to high temperatures (millions of kelvin).

If the gas or liquid is very viscous (resulting in turbulent differential motion), the reversal of the magnetic field may not be very periodic. This is the case with the Earth's magnetic field, which is generated by turbulent currents in a viscous outer core.

Surface Activity

Starspots are regions of intense magnetic activity on the surface of a star. (On the Sun they are termed sunspots.) These form a visible component of magnetic flux tubes that are formed within a star's convection zone. Due to the differential rotation of the star, the tube becomes curled up and stretched, inhibiting convection and producing zones of lower than normal temperature. Coronal loops often form above starspots, forming from magnetic field lines that stretch out into the corona. These in turn serve to heat the corona to temperatures over a million kelvins.

The magnetic fields linked to starspots and coronal loops are linked to flare activity, and the associated coronal mass ejection. The plasma is heated to tens of millions of kelvins, and the particles are accelerated away from the star's surface at extreme velocities.

Surface activity appears to be related to the age and rotation rate of main-sequence stars. Young stars with a rapid rate of rotation exhibit strong activity.

By contrast middle-aged, Sun-like stars with a slow rate of rotation show low levels of activity that varies in cycles. Some older stars display almost no activity, which may mean they have entered a lull that is comparable to the Sun's Maunder minimum. Measurements of the time variation in stellar activity can be useful for determining the differential rotation rates of a star.

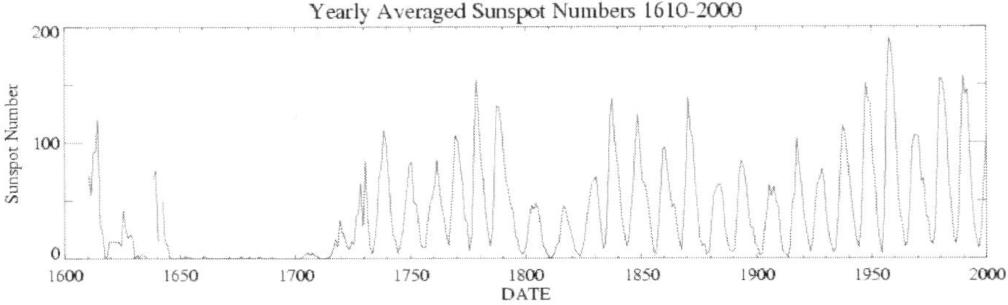

Magnetosphere

A star with a magnetic field will generate a magnetosphere that extends outward into the surrounding space. Field lines from this field originate at one magnetic pole on the star then end at the other pole, forming a closed loop. The magnetosphere contains charged particles that are trapped from the stellar wind, which then move along these field lines. As the star rotates, the magnetosphere rotates with it, dragging along the charged particles.

As stars emit matter with a stellar wind from the photosphere, the magnetosphere creates a torque on the ejected matter. This results in a transfer of angular momentum from the star to the surrounding space, causing a slowing of the stellar rotation rate. Rapidly rotating stars have a higher mass loss rate, resulting in a faster loss of momentum. As the rotation rate slows, so too does the angular deceleration. By this means, a star will gradually approach, but never quite reach, the state of zero rotation.

Magnetic Stars

Figure: *Surface magnetic field of SU Aur (a young star of T Tauri type), reconstructed by means of Zeeman-Doppler imaging*

A T Tauri star is a type of pre–main sequence star that is being heated through gravitational contraction and has not yet begun to burn hydrogen at its core. They are variable stars that are magnetically active. The magnetic field of these stars is thought to interact with its strong stellar wind, transferring angular momentum to the surrounding protoplanetary disk. This allows the star to brake its rotation rate as it collapses.

Small, M-class stars (with 0.1–0.6 solar masses) that exhibit rapid, irregular variability are known as flare stars. These fluctuations are hypothesized to be caused by flares, although the activity is much stronger relative to the size of the star. The flares on this class of stars can extend up to 20% of the circumference, and radiate much of their energy in the blue and ultraviolet portion of the spectrum.

Planetary nebulae are created when a red giant star ejects its outer envelope, forming an expanding shell of gas. However it remains a mystery why these shells are not always spherically symmetrical. 80%

of planetary nebulae do not have a spherical shape; instead forming bipolar or elliptical nebulae. One hypothesis for the formation of a non-spherical shape is the effect of the star's magnetic field. Instead of expanding evenly in all directions, the ejected plasma tends to leave by way of the magnetic poles. Observations of the central stars in at least four planetary nebulae have confirmed that they do indeed possess powerful magnetic fields.

After some massive stars have ceased thermonuclear fusion, a portion of their mass collapses into a compact body of neutrons called a neutron star. These bodies retain a significant magnetic field from the original star, but the collapse in size causes the strength of this field to increase dramatically. The rapid rotation of these collapsed neutron stars results in a pulsar, which emits a narrow beam of energy that can periodically point toward an observer.

Compact and fast-rotating astronomical objects (white dwarfs, neutron stars and black holes) have extremely strong magnetic fields. The magnetic field of a newly born fast-spinning neutron star is so strong (up to 10^8 teslas) that it electromagnetically radiates enough energy to quickly (in a matter of few million years) damp down the star rotation by 100 to 1000 times. Matter falling on a neutron star also has to follow the magnetic field lines, resulting in two hot spots on the surface where it can reach and collide with the star's surface. These spots are literally a few feet (about a metre) across but tremendously bright. Their periodic eclipsing during star rotation is hypothesized to be the source of pulsating radiation.

An extreme form of a magnetized neutron star is the magnetar. These are formed as the result of a core-collapse supernova. The existence of such stars was confirmed in 1998 with the measurement of the star SGR 1806-20. The magnetic field of this star has increased the surface temperature to 18 million K and it releases enormous amounts of energy in gamma ray bursts.

Jets of relativistic plasma are often observed along the direction of the magnetic poles of active black holes in the centers of very young galaxies.

Pulsar

A pulsar (short for *pulsating radio star*) is a highly magnetized, rotating neutron star that emits a beam of electromagnetic radiation. This radiation can only be observed when the beam of emission is pointing toward the Earth, much the way a lighthouse can only be seen when the light is pointed in the direction of an observer, and is responsible for the pulsed appearance of emission. Neutron stars are very dense, and have short, regular rotational periods. This produces a very precise interval between pulses that range from roughly milliseconds to seconds for an individual pulsar.

The precise periods of pulsars make them useful tools. Observations of a pulsar in a binary neutron star system were used to indirectly confirm the existence of gravitational radiation. The first extrasolar planets were discovered around a pulsar, PSR B1257+12. Certain types of pulsars rival atomic clocks in their accuracy in keeping time.

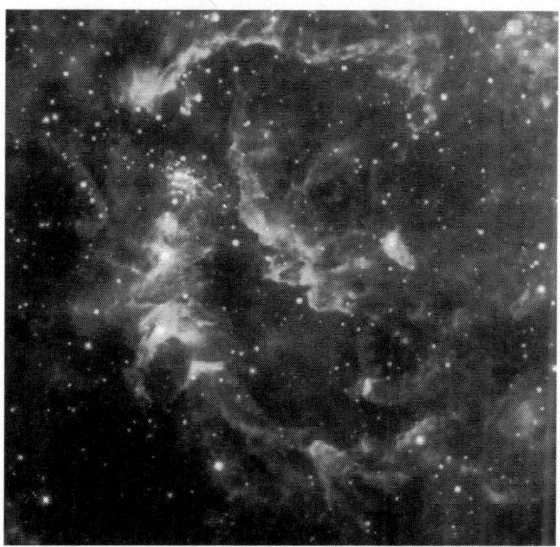

Figure: *PSR B1509-58 - X-rays from Chandra are gold; Infrared from WISE in red, green and blue/max.*

History of Observation

Discovery

The first pulsar was observed on November 28, 1967, by Jocelyn Bell Burnell and Antony Hewish. They observed pulses separated by 1.33 seconds that originated from the same location on the sky, and kept to sidereal time. In looking for explanations for the pulses, the short period of the pulses eliminated most astrophysical sources of radiation, such as stars, and since the pulses followed sidereal time, it could not be man-made radio frequency interference. When observations with another telescope confirmed the emission, it eliminated any sort of instrumental effects. At this point, Burnell notes of herself and Hewish that "we did not really believe that we had picked up signals from another civilization, but obviously the idea had crossed our minds and we had no proof that it was an entirely natural radio emission. It is an interesting problem—if one thinks one may have detected life elsewhere in the universe, how does one announce the results responsibly?"

Even so, they nicknamed the signal LGM-1, for "little green men" (a playful name for intelligent beings of extraterrestrial origin). It was not until a second pulsating source was discovered in a different part of the sky that the "LGM hypothesis" was entirely abandoned.

Figure: *Composite Optical/X-ray image of the Crab Nebula, showing synchrotron emission in the surrounding pulsar wind nebula, powered by injection of magnetic fields and particles from the central pulsar.*

Their pulsar was later dubbed CP 1919, and is now known by a number of designators including PSR 1919+21, PSR B1919+21 and PSR J1921+2153. Although CP 1919 emits in radio wavelengths, pulsars have, subsequently, been found to emit in visible light, X-ray, and/or gamma ray wavelengths.

The word "pulsar" is a contraction of "pulsating star", and first appeared in print in 1968:

> *An entirely novel kind of star came to light on Aug. 6 last year and was referred to, by astronomers, as LGM (Little Green Men). Now it is thought to be a novel type between a white dwarf and a neutron [*sic*]. The name Pulsar is likely to be given to it. Dr. A. Hewish told me yesterday: "... I am sure that today every radio telescope is looking at the Pulsars."*

The existence of neutron stars was first proposed by Walter Baade and Fritz Zwicky in 1934, when they argued that a small, dense star consisting primarily of neutrons would result from a supernova. In 1967, shortly before the discovery of pulsars, Franco Pacini suggested that a rotating neutron star with a magnetic field would emit radiation, and even noted that such energy could be pumped into a supernova remnant around a neutron star, such as the Crab Nebula.

After the discovery of the first pulsar, Thomas Gold independently suggested a rotating neutron star model similar to that of Pacini, and explicitly argued that this model could explain the pulsed radiation observed by Bell Burnell and Hewish. The discovery of the Crab pulsar later in 1968 seemed to provide confirmation of the rotating neutron star model of pulsars.

The Crab pulsar has a 33-millisecond pulse period, which was too short to be consistent with other proposed models for pulsar emission. A rotation speed of 1,980 revolutions per minute was considered perfectly acceptable. Moreover, the Crab pulsar is so named because it is located at the center of the Crab Nebula, consistent with the 1933 prediction of Baade and Zwicky.

In 1974, Antony Hewish and Martin Ryle became the first astronomers to be awarded the Nobel Prize in physics, with the Royal Swedish Academy of Sciences noting that Hewish played a "decisive role in the discovery of pulsars".

Considerable controversy is associated with the fact that Professor Hewish was awarded the prize while Bell, who made the initial discovery while she was his Ph.D student, was not. Bell claims no bitterness upon this point, supporting the decision of the Nobel prize committee.

Milestones

In 1974, Joseph Hooton Taylor, Jr. and Russell Hulse discovered for the first time a pulsar in a binary system, PSR B1913+16. This pulsar orbits another neutron star with an orbital period of just eight hours. Einstein's theory of general relativity predicts that this system should emit strong gravitational radiation, causing the orbit to continually contract as it loses orbital energy. Observations of the pulsar soon confirmed this prediction, providing the first ever evidence of the existence of gravitational waves.

As of 2010, observations of this pulsar continue to agree with general relativity. In 1993, the Nobel Prize in Physics was awarded to Taylor and Hulse for the discovery of this pulsar.

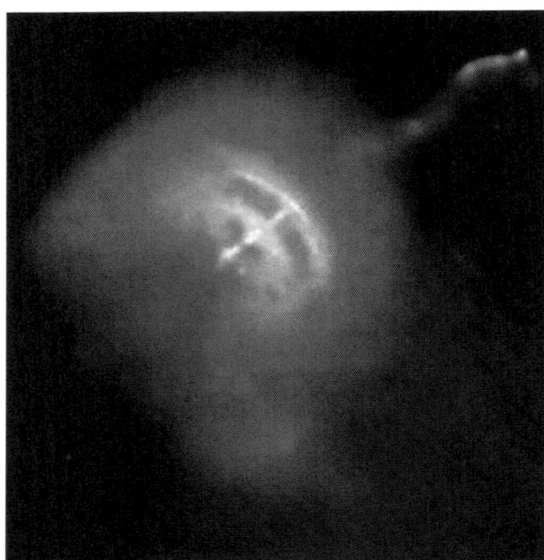

Figure: The Vela Pulsar and its surrounding pulsar wind nebula.

In 1982, Don Backer led a group which discovered PSR B1937+21, a pulsar with a rotation period of just 1.6 milliseconds (38,500 rpm). Observations soon revealed that its magnetic field was much weaker than ordinary pulsars, while further discoveries cemented the idea that a new class of object, the "millisecond pulsars" (MSPs) had been found. MSPs are believed to be the end product of X-ray binaries. Owing to their extraordinarily rapid and stable rotation, MSPs can be used by astronomers as clocks rivaling the stability of the best atomic clocks on Earth. Factors affecting the arrival time of pulses at the Earth by more than a few hundred nanoseconds can be easily detected and used to make precise measurements.

Physical parameters accessible through pulsar timing include the 3D position of the pulsar, its proper motion, the electron content of the interstellar medium along the propagation path, the orbital parameters of any binary companion, the pulsar rotation period and its evolution with time. (These are computed from the raw timing data by Tempo, a computer program specialised for this task.)

After these factors have been taken into account, deviations between the observed arrival times and predictions made using these parameters can be found and attributed to one of three possibilities: intrinsic variations in the spin period of the pulsar, errors in the realisation of Terrestrial Time against which arrival times were measured, or the presence of background gravitational waves. Scientists are currently attempting to resolve these possibilities by comparing the deviations seen between several different pulsars, forming what is known as a Pulsar timing array. The goal of these efforts is to develop a pulsar-based time standard precise enough to make the first ever direct detection of gravitational waves. In June 2006, the astronomer John Middleditch and his team at LANL announced the first prediction of pulsar glitches with observational data from the Rossi X-ray Timing Explorer. They used observations of the pulsar PSR J0537-6910.

In 1992, Aleksander Wolszczan discovered the first extrasolar planets around PSR B1257+12. This discovery presented important evidence concerning the widespread existence of planets outside the solar system, although it is very unlikely that any life form could survive in the environment of intense radiation near a pulsar.

Nomenclature

Initially pulsars were named with letters of the discovering observatory followed by their right ascension (e.g. CP 1919). As more pulsars were discovered, the letter code became unwieldy, and so the convention then arose of using the letters PSR (Pulsating Source of Radio) followed by the pulsar's right ascension and degrees of declination (e.g. PSR 0531+21) and sometimes declination to a tenth of a degree (e.g. PSR 1913+167). Pulsars appearing very close together sometimes have letters appended (e.g. PSR 0021-72C and PSR 0021-72D).

The modern convention prefixes the older numbers with a B (e.g. PSR B1919+21), with the B meaning the coordinates are for the 1950.0 epoch. All new pulsars have a J indicating 2000.0 coordinates and also have declination including minutes (e.g. PSR J1921+2153). Pulsars that were discovered before 1993 tend to retain their B names rather than use their J names (e.g. PSR J1921+2153 is more commonly known as PSR B1919+21). Recently discovered pulsars only have a J name (e.g. PSR J0437-4715). All pulsars have a J name that provides more precise coordinates of its location in the sky.

Formation

The events leading to the formation of a pulsar begin when the core of a massive star is compressed during a supernova, which collapses into a neutron star. The neutron star retains most of its angular momentum, and since it has only a tiny fraction of its progenitor's radius (and therefore its moment of inertia is sharply reduced), it is formed with very high rotation speed. A beam of radiation is emitted along the magnetic axis of the pulsar, which spins along with the rotation of the neutron star. The magnetic axis of the pulsar determines the direction of the electromagnetic beam, with the magnetic axis not necessarily being the same as its rotational axis.

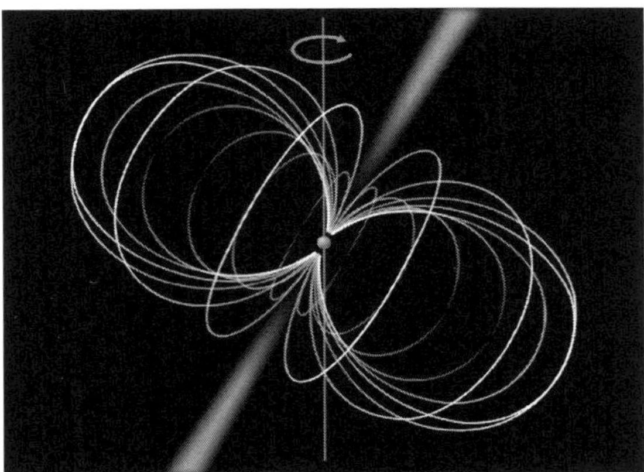

Figure: *Schematic view of a pulsar. The sphere in the middle represents the neutron star, the curves indicate the magnetic field lines, the protruding cones represent the emission beams and the green line represents the axis on which the star rotates.*

This misalignment causes the beam to be seen once for every rotation of the neutron star, which leads to the "pulsed" nature of its appearance. The beam originates from the rotational energy of the neutron star, which generates an electrical field from the movement of the very strong magnetic field, resulting in the acceleration of protons and electrons on the star surface and the creation of an electromagnetic beam emanating from the poles of the magnetic field.

This rotation slows down over time as electromagnetic power is emitted. When a pulsar's spin period slows down sufficiently, the radio pulsar mechanism is believed to turn off (the so-called "death line"). This turn-off seems to take place after about 10–100 million years, which means of all the neutron stars in the 13.6 billion year age of the universe, around 99% no longer pulsate. The longest known pulsar period is 8.51 seconds.

Though the general picture of pulsars as rapidly rotating neutron stars is widely accepted, Werner Becker of the Max Planck Institute for Extraterrestrial Physics said in 2006, "The theory of how pulsars emit their radiation is still in its infancy, even after nearly forty years of work."

Categories

Three distinct classes of pulsars are currently known to astronomers, according to the source of the power of the electromagnetic radiation:

- Rotation-powered pulsars, where the loss of rotational energy of the star provides the power.

- Accretion-powered pulsars (accounting for most but not all X-ray pulsars), where the gravitational potential energy of accreted matter is the power source (producing X-rays that are observable from the Earth).
- Magnetars, where the decay of an extremely strong magnetic field provides the electromagnetic power.

The Fermi Space Telescope has uncovered a subclass of rotationally-powered pulsars that emit only gamma rays. There have been only about one hundred gamma-ray pulsars identified out of about 1800 known pulsars.

Although all three classes of objects are neutron stars, their observable behaviour and the underlying physics are quite different. There are, however, connections. For example, X-ray pulsars are probably old rotationally-powered pulsars that have already lost most of their power, and have only become visible again after their binary companions had expanded and began transferring matter on to the neutron star. The process of accretion can in turn transfer enough angular momentum to the neutron star to "recycle" it as a rotation-powered millisecond pulsar. As this matter lands on the neutron star, it is thought to "bury" the magnetic field of the neutron star (although the details are unclear), leaving millisecond pulsars with magnetic fields 1000-10,000 times weaker than average pulsars. This low magnetic field is less effective at slowing the pulsar's rotation, so millisecond pulsars live for billions of years, making them the oldest known pulsars. Millisecond pulsars are seen in globular clusters, which stopped forming neutron stars billions of years ago.

Of interest to the study of the state of the matter in a neutron stars are the *glitches* observed in the rotation velocity of the neutron star. This velocity is decreasing slowly but steadily, except by sudden variations. One model put forward to explain these glitches is that they are the result of "starquakes" that adjust the crust of the neutron star. Models where the glitch is due to a decoupling of the possibly superconducting interior of the star have also been advanced. In both cases, the star's moment of inertia changes, but its angular momentum does not, resulting in a change in rotation rate.

Disrupted Recycled Pulsar

When two massive stars are born close together from the same cloud of gas, they can form a binary system and orbit each other from birth. If those two stars are at least a few times as massive as our sun, their lives will both end in supernova explosions. The more massive star explodes first, leaving behind a neutron star. If the explosion does not kick the second star away, the binary system survives. The neutron star can now be visible as a radio pulsar, and it slowly loses energy and spins down. Later, the second star can swell up, allowing the neutron star to suck up its matter. T

he matter falling onto the neutron star spins it up and reduces its magnetic field. This is called "recycling" because it returns the neutron star to a quickly-spinning state. Finally, the second star also explodes in a supernova, producing another neutron star. If this second explosion also fails to disrupt the binary, a double neutron star binary is formed. Otherwise, the spun-up neutron star is left with no companion and becomes a "disrupted recycled pulsar", spinning between a few and 50 times per second.

Applications

The discovery of pulsars allowed astronomers to study an object never observed before, the neutron star. This kind of object is the only place where the behaviour of matter at nuclear density can be observed (though not directly). Also, millisecond pulsars have allowed a test of general relativity in conditions of an intense gravitational field.

Maps

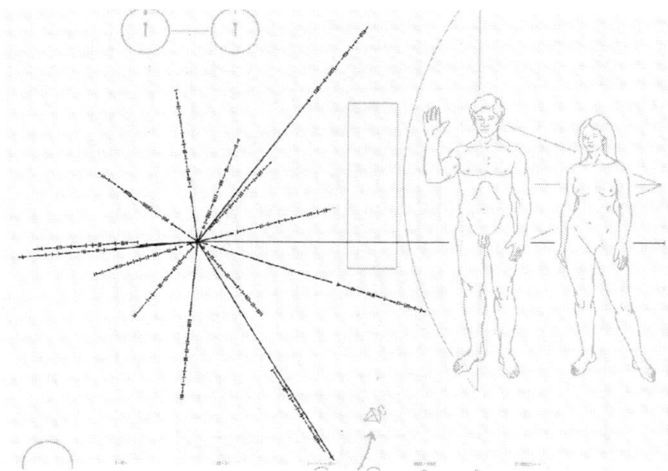

Figure: *Relative position of the Sun to the center of the Galaxy and 14 pulsars with their periods denoted*

Pulsar maps have been included on the two Pioneer Plaques as well as the Voyager Golden Record. They show the position of the Sun, relative to 14 pulsars, which are identified by the unique timing of their electromagnetic pulses, so that our position both in space and in time can be calculated by potential extraterrestrial intelligences. Because pulsars are emitting very regular pulses of radio waves, its radio transmissions do not require daily corrections. Moreover, pulsar positioning could create a spacecraft navigation system independently, or be an auxiliary device to GPS instruments.

Precise Clocks

For some millisecond pulsars, the regularity of pulsation is more precise than an atomic clock. This stability allows millisecond pulsars to be used in establishing ephemeris time or building pulsar clocks.

Timing noise is the name for rotational irregularities observed in all pulsars. This timing noise is observable as random wandering in the pulse frequency or phase. It is unknown whether timing noise is related to pulsar glitches.

Probes of the Interstellar Medium

The radiation from pulsars passes through the interstellar medium (ISM) before reaching Earth. Free electrons in the warm (8000 K), ionized component of the ISM and H II regions affect the radiation in two primary ways. The resulting changes to the pulsar's radiation provide an important probe of the ISM itself.

Because of the dispersive nature of the interstellar plasma, lower-frequency radio waves travel through the medium slower than higher-frequency radio waves. The resulting delay in the arrival of pulses at a range of frequencies is directly measurable as the *dispersion measure* of the pulsar. The dispersion measure is the total column density of free electrons between the observer and the pulsar,

$$\text{DM} = \int_0^D n_e(s)ds,$$

where D is the distance from the pulsar to the observer and n_e is the electron density of the ISM. The dispersion measure is used to construct models of the free electron distribution in the Milky Way Galaxy.

Additionally, turbulence in the interstellar gas causes density inhomogeneities in the ISM which cause scattering of the radio waves from the pulsar. The resulting scintillation of the radio waves—the same effect as the twinkling of a star in visible light due to density variations in the Earth's atmosphere—can be used to reconstruct information about the small scale variations in the ISM. Due to the high velocity (up to several hundred km/s) of many pulsars, a single pulsar scans the ISM rapidly, which results in changing scintillation patterns over timescales of a few minutes.

Probes of Space-Time

Gravitational Waves Detectors

There are 3 consortia around the world which use pulsars to search for gravitational waves. In Europe, there is the European Pulsar Timing Array (EPTA); there is the Parkes Pulsar Timing Array (PPTA) in Australia; and there is the North American Nanohertz Observatory for Gravitational Waves (NANOGrav) in Canada and the US. Together, the consortia form the International Pulsar Timing Array (IPTA). The pulses from Millisecond Pulsars (MSPs) are used as a system of Galactic clocks. Disturbances in the clocks will be measurable at Earth. A disturbance from a passing gravitational wave will have a particular signature across the ensemble of pulsars, and will be thus detected.

Significant Pulsars

Pulsars within 300 pc		
PSRJ	*Distance (pc)*	*Age (Myr)*
J0030+0451	244	7,580
0108–1431	238	166
0437–4715	156	1,590
0633+1746	156	0.342
0659+1414	290	0.111
0835–4510	290	0.0113
0453+0755	260	17.5
1045–4509	300	6,710
1741–2054	250	0.387
1856–3754	161	3.76
2144–3933	165	272

The pulsars listed here were either the first discovered of its type, or represent an extreme of some type among the known pulsar population, such as having the shortest measured period.

- The first radio pulsar "CP 1919" (now known as PSR B1919+21), with a pulse period of 1.337 seconds and a pulse width of 0.04 second, was discovered in 1967.
- The first binary pulsar, PSR 1913+16, whose orbit is decaying at the exact rate predicted due to the emission of gravitational radiation by general relativity
- The first millisecond pulsar, PSR B1937+21
- The brightest millisecond pulsar, PSR J0437-4715
- The first X-ray pulsar, Cen X-3

- The first accreting millisecond X-ray pulsar, SAX J1808.4-3658
- The first pulsar with planets, PSR B1257+12
- The first pulsar observed to have been affected by asteroids: PSR J0738-4042
- The first double pulsar binary system, PSR J0737d3039
- The longest period pulsar, PSR J2144-3933
- The most stable pulsar in period, PSR J0437-4715
- The first millisecond pulsar with 2 stellar mass companions, PSR J0337+1715
- PSR B1931+24 "... appears as a normal pulsar for about a week and then 'switches off' for about one month before emitting pulses again. [..] this pulsar slows down more rapidly when the pulsar is on than when it is off. [.. the] braking mechanism must be related to the radio emission and the processes creating it and the additional slow-down can be explained by the pulsar wind leaving the pulsar's magnetosphere and carrying away rotational energy."
- PSR J1748-2446ad, at 716 Hz, the pulsar with the highest rotation speed.
- PSR J1903+0327, a ~2.15 ms pulsar discovered to be in a highly eccentric binary star system with a sun-like star.
- A pulsar in the CTA 1 supernova remnant (4U 0000+72, in Cassiopeia) was found by the Fermi Gamma-ray Space Telescope to emit pulsations only in gamma ray radiation, the first recorded of its kind.
- PSR J2007+2722, a 40.8-hertz 'recycled' isolated pulsar was the first pulsar found by volunteers on data taken in February 2007 and analyzed by distributed computing project Einstein@Home.
- PSR J1311–3430, the first millisecond pulsar discovered via gamma-ray pulsations and part of a binary system with the shortest orbital period.

Black Hole

A black hole is a region of spacetime from which gravity prevents anything, including light, from escaping. The theory of general relativity predicts that a sufficiently compact mass will deform spacetime to form a black hole. The boundary of the region from which no escape is possible is called the event horizon. Although crossing the event horizon has enormous effect on the fate of the object crossing it, it appears to have no locally detectable features. In many ways a black hole acts like an ideal black body, as it reflects no light. Moreover, quantum field theory in curved spacetime predicts that event horizons emit Hawking radiation, with the same spectrum as a black body of a temperature inversely proportional to its mass. This temperature is on the order of billionths of a kelvin for black holes of stellar mass, making it all but impossible to observe.

Figure: *Simulated view of a black hole (center) in front of the Large Magellanic Cloud. Note the gravitational lensing effect, which produces two enlarged but highly distorted views of the Cloud. Across the top, the Milky Way disk appears distorted into an arc.*

Objects whose gravitational fields are too strong for light to escape were first considered in the 18th century by John Michell and Pierre-Simon Laplace. The first modern solution of general relativity that would characterize a black hole was found by Karl Schwarzschild in 1916, although its interpretation as a region of space from which nothing can escape was first published by David Finkelstein in 1958. Long considered a mathematical curiosity, it was during the 1960s that theoretical work showed black holes were a generic prediction of general relativity. The discovery of neutron stars sparked interest in gravitationally collapsed compact objects as a possible astrophysical reality.

Black holes of stellar mass are expected to form when very massive stars collapse at the end of their life cycle. After a black hole has formed, it can continue to grow by absorbing mass from its surroundings. By absorbing other stars and merging with other black holes, supermassive black holes of millions of solar masses may form. There is general consensus that supermassive black holes exist in the centers of most galaxies.

Despite its invisible interior, the presence of a black hole can be inferred through its interaction with other matter and with electromagnetic radiation such as light. Matter falling onto a black hole can form an accretion disk heated by friction, forming some of the brightest objects in the universe. If there are other stars orbiting a black hole, their orbit can be used to determine its mass and location. Such observations can be used to exclude possible alternatives (such as neutron stars). In this way, astronomers have identified numerous stellar black hole candidates in binary systems, and established that the core of the Milky Way contains a supermassive black hole of about 4.3 million solar masses.

History

The idea of a body so massive that even light could not escape was first put forward by John Michell in a letter written to Henry Cavendish in 1783 of the Royal Society:

> *If the semi-diameter of a sphere of the same density as the Sun were to exceed that of the Sun in the proportion of 500 to 1, a body falling from an infinite height towards it would have acquired at its surface greater velocity than that of light, and consequently supposing light to be attracted by the same force in proportion to its vis inertiae, with other bodies, all light emitted from such a body would be made to return towards it by its own proper gravity.* —*John Michell*

In 1796, mathematician Pierre-Simon Laplace promoted the same idea in the first and second editions of his book *Exposition du système du Monde* (it was removed from later editions). Such "dark stars" were largely ignored in the nineteenth century, since it was not understood how a massless wave such as light could be influenced by gravity.

General Relativity

In 1915, Albert Einstein developed his theory of general relativity, having earlier shown that gravity does influence light's motion. Only a few months later, Karl Schwarzschild found a solution to the Einstein field equations, which describes the gravitational field of a point mass and a spherical mass. A few months after Schwarzschild, Johannes Droste, a student of Hendrik Lorentz, independently gave the same solution for the point mass and wrote more extensively about its properties. This solution had a peculiar behaviour at what is now called the Schwarzschild radius, where it became singular, meaning that some of the terms in the Einstein equations became infinite. The nature of this surface was not quite understood at the time. In 1924, Arthur Eddington showed that the singularity disappeared after a change of coordinates, although it took until 1933 for Georges Lemaître to realise that this meant the singularity at the Schwarzschild radius was an unphysical coordinate singularity.

In 1931, Subrahmanyan Chandrasekhar calculated, using special relativity, that a non-rotating body of electron-degenerate matter above a certain limiting mass (now called the Chandrasekhar limit at 1.4 solar masses) has no stable solutions. His arguments were opposed by many of his contemporaries like Eddington and Lev Landau, who argued that some yet unknown mechanism would stop the collapse. They were partly correct: a white dwarf slightly more massive than the Chandrasekhar limit will collapse into a neutron star, which is itself stable because of the Pauli exclusion principle. But in 1939, Robert Oppenheimer and others predicted that neutron stars above approximately three solar masses (the Tolman–Oppenheimer–Volkoff limit) would collapse into black holes for the reasons presented by Chandrasekhar, and concluded that no law of physics was likely to intervene and stop at least some stars from collapsing to black holes.

Oppenheimer and his co-authors interpreted the singularity at the boundary of the Schwarzschild radius as indicating that this was the boundary of a bubble in which time stopped. This is a valid point of view for external observers, but not for infalling observers. Because of this property, the collapsed stars were called "frozen stars", because an outside observer would see the surface of the star frozen in time at the instant where its collapse takes it inside the Schwarzschild radius.

Golden Age

In 1958, David Finkelstein identified the Schwarzschild surface as an event horizon, "a perfect unidirectional membrane: causal influences can cross it in only one direction". This did not strictly contradict Oppenheimer's results, but extended them to include the point of view of infalling observers. Finkelstein's solution extended the Schwarzschild solution for the future of observers falling into a black hole. A complete extension had already been found by Martin Kruskal, who was urged to publish it.

These results came at the beginning of the golden age of general relativity, which was marked by general relativity and black holes becoming mainstream subjects of research. This process was helped by the discovery of pulsars in 1967, which, by 1969, were shown to be rapidly rotating neutron stars. Until that time, neutron stars, like black holes, were regarded as just theoretical curiosities; but the discovery of pulsars showed their physical relevance and spurred a further interest in all types of compact objects that might be formed by gravitational collapse.

In this period more general black hole solutions were found. In 1963, Roy Kerr found the exact solution for a rotating black hole. Two years later, Ezra Newman found the axisymmetric solution for a black hole that is both rotating and electrically charged. Through the work of Werner Israel, Brandon Carter, and David Robinson the no-hair theorem emerged, stating that a stationary black hole solution is completely described by the three parameters of the Kerr–Newman metric; mass, angular momentum, and electric charge.

At first, it was suspected that the strange features of the black hole solutions were pathological artifacts from the symmetry conditions imposed, and that the singularities would not appear in generic situations. This view was held in particular by Vladimir Belinsky, Isaak Khalatnikov, and Evgeny Lifshitz, who tried to prove that no singularities appear in generic solutions. However, in the late 1960s Roger Penrose and Stephen Hawking used global techniques to prove that singularities appear generically.

Work by James Bardeen, Jacob Bekenstein, Carter, and Hawking in the early 1970s led to the formulation of black hole thermodynamics. These laws describe the behaviour of a black hole in close analogy to the laws of thermodynamics by relating mass to energy, area to entropy, and surface gravity to temperature. The analogy was completed when Hawking, in 1974, showed that quantum field theory predicts that black holes should radiate like a black body with a temperature proportional to the surface gravity of the black hole.

The first use of the term "black hole" in print was by journalist Ann Ewing in her article *"'Black Holes' in Space"*, dated 18 January 1964, which was a report on a meeting of the American Association for the Advancement of Science. John Wheeler used term "black hole" a lecture in 1967, leading some to credit him with coining the phrase. After Wheeler's use of the term, it was quickly adopted in general use.

Properties and Structure

The no-hair theorem states that, once it achieves a stable condition after formation, a black hole has only three independent physical properties: mass, charge, and angular momentum. Any two black holes that share the same values for these properties, or parameters, are indistinguishable according to classical (i.e. non-quantum) mechanics.

These properties are special because they are visible from outside a black hole. For example, a charged black hole repels other like charges just like any other charged object. Similarly, the total mass inside a sphere containing a black hole can be found by using the gravitational analog of Gauss's law, the ADM mass, far away from the black hole. Likewise, the angular momentum can be measured from far away using frame dragging by the gravitomagnetic field.

When an object falls into a black hole, any information about the shape of the object or distribution of charge on it is evenly distributed along the horizon of the black hole, and is lost to outside observers. The behaviour of the horizon in this situation is a dissipative system that is closely analogous to that of a conductive stretchy membrane with friction and electrical resistance—the membrane paradigm. This is different from other field theories like electromagnetism, which do not have any friction or resistivity at the microscopic level, because they are time-reversible. Because a black hole eventually achieves a stable state with only three parameters, there is no way to avoid losing information about the initial conditions: the gravitational and electric fields of a black hole give very little information about what went in. The information that is lost includes every quantity that cannot be measured far away from the black hole horizon, including approximately conserved quantum numbers such as the total baryon number and lepton number. This behaviour is so puzzling that it has been called the black hole information loss paradox.

Physical Properties

The simplest static black holes have mass but neither electric charge nor angular momentum. These black holes are often referred to as Schwarzschild black holes after Karl Schwarzschild who discovered this solution in 1916. According to Birkhoff's theorem, it is the only vacuum solution that is spherically symmetric. This means that there is no observable difference between the gravitational field of such a black hole and that of any other spherical object of the same mass. The popular notion of a black hole "sucking in everything" in its surroundings is therefore only correct near a black hole's horizon; far away, the external gravitational field is identical to that of any other body of the same mass.

Solutions describing more general black holes also exist. Charged black holes are described by the Reissner–Nordstrφm metric, while the Kerr metric describes a rotating black hole. The most general stationary black hole solution known is the Kerr–Newman metric, which describes a black hole with both charge and angular momentum. While the mass of a black hole can take any positive value, the charge and angular momentum are constrained by the mass. In Planck units, the total electric charge Q and the total angular momentum J are expected to satisfy

$$Q^2 + \left(\tfrac{J}{M}\right)^2 \leq M^2$$

for a black hole of mass M. Black holes saturating this inequality are called extremal. Solutions of Einstein's equations that violate this inequality exist, but they do not possess an event horizon. These solutions have so-called naked singularities that can be observed from the outside, and hence are deemed *unphysical*. The cosmic censorship hypothesis rules out the formation of such singularities, when they are created through the gravitational collapse of realistic matter. This is supported by numerical simulations.

Due to the relatively large strength of the electromagnetic force, black holes forming from the collapse of stars are expected to retain the nearly neutral charge of the star. Rotation, however, is expected to be a common feature of compact objects. The black-hole candidate binary X-ray source GRS 1915+105 appears to have an angular momentum near the maximum allowed value.

Black hole classifications		
Class	***Mass***	***Size***
Supermassive black hole	$\sim 10^5$–10^{10} M_{Sun}	~0.001–400 AU
Intermediate-mass black hole	$\sim 10^3$ M_{Sun}	$\sim 10^3$ km $\approx R_{Earth}$
Stellar black hole	~ 10 M_{Sun}	~30 km
Micro black hole	up to $\sim M_{Moon}$	up to ~0.1 mm

Black holes are commonly classified according to their mass, independent of angular momentum J or electric charge Q. The size of a black hole, as determined by the radius of the event horizon, or Schwarzschild radius, is roughly proportional to the mass M through

$$r_{sh} = \frac{2GM}{c^2} \approx 2.95 \frac{M}{M_{Sun}} \text{km},$$

where r_{sh} is the Schwarzschild radius and M_{Sun} is the mass of the Sun. This relation is exact only for black holes with zero charge and angular momentum; for more general black holes it can differ up to a factor of 2.

Event Horizon

The defining feature of a black hole is the appearance of an event horizon—a boundary in spacetime through which matter and light can only pass inward towards the mass of the black hole. Nothing, not even light, can escape from inside the event horizon. The event horizon is referred to as such because if an event occurs within the boundary, information from that event cannot reach an outside observer, making it impossible to determine if such an event occurred.

As predicted by general relativity, the presence of a mass deforms spacetime in such a way that the paths taken by particles bend towards the mass. At the event horizon of a black hole, this deformation becomes so strong that there are no paths that lead away from the black hole.

To a distant observer, clocks near a black hole appear to tick more slowly than those further away from the black hole. Due to this effect, known as gravitational time dilation, an object falling into a black hole appears to slow down as it approaches the event horizon, taking an infinite time to reach it. At the same time, all processes on this object slow down, for a fixed outside observer, causing emitted light to appear redder and dimmer, an effect known as gravitational redshift. Eventually, at a point just before it reaches the event horizon, the falling object becomes so dim that it can no longer be seen.

On the other hand, an indestructible observer falling into a black hole does not notice any of these effects as he crosses the event horizon. According to his own clock, which appears to him to tick normally, he crosses the event horizon after a finite time without noting any singular behaviour. In particular, he is unable to determine exactly when he crosses it, as it is impossible to determine the location of the event horizon from local observations.

The shape of the event horizon of a black hole is always approximately spherical. For non-rotating (static) black holes the geometry is precisely spherical, while for rotating black holes the sphere is somewhat oblate.

Singularity

At the center of a black hole as described by general relativity lies a gravitational singularity, a region where the spacetime curvature becomes infinite. For a non-rotating black hole, this region takes the shape of a single point and for a rotating black hole, it is smeared out to form a ring singularity lying in the plane of rotation. In both cases, the singular region has zero volume. It can also be shown that the singular region contains all the mass of the black hole solution. The singular region can thus be thought of as having infinite density.

Observers falling into a Schwarzschild black hole (*i.e.*, non-rotating and not charged) cannot avoid being carried into the singularity, once they cross the event horizon. They can prolong the experience by accelerating away to slow their descent, but only up to a point; after attaining a certain ideal velocity, it is best to free fall the rest of the way. When they reach the singularity, they are crushed to infinite density and their mass is added to the total of the black hole. Before that happens, they will have been torn apart by the growing tidal forces in a process sometimes referred to as spaghettification or the "noodle effect".

In the case of a charged (Reissner–Nordström) or rotating (Kerr) black hole, it is possible to avoid the singularity. Extending these solutions as far as possible reveals the hypothetical possibility of exiting the black hole into a different spacetime with the black hole acting as a wormhole. The possibility of travelling to another universe is however only theoretical, since any perturbation will destroy this possibility. It also appears to be possible to follow closed timelike curves (going back to one's own past) around the Kerr singularity, which lead to problems with causality like the grandfather paradox. It is expected that none of these peculiar effects would survive in a proper quantum treatment of rotating and charged black holes.

The appearance of singularities in general relativity is commonly perceived as signaling the breakdown of the theory. This breakdown, however, is expected; it occurs in a situation where quantum effects should describe these actions, due to the extremely high density and therefore particle interactions. To date, it has not been possible to combine quantum and gravitational effects into a single theory, although there exist attempts to formulate such a theory of quantum gravity. It is generally expected that such a theory will not feature any singularities.

Photon Sphere

The photon sphere is a spherical boundary of zero thickness such that photons moving along tangents to the sphere will be trapped in a circular orbit. For non-rotating black holes, the photon sphere has a radius 1.5 times the Schwarzschild radius. The orbits are dynamically unstable, hence any small perturbation (such as a particle of infalling matter) will grow over time, either setting it on an outward trajectory escaping the black hole or on an inward spiral eventually crossing the event horizon.

While light can still escape from inside the photon sphere, any light that crosses the photon sphere on an inbound trajectory will be captured by the black hole. Hence any light reaching an outside observer from inside the photon sphere must have been emitted by objects inside the photon sphere but still outside of the event horizon.

Other compact objects, such as neutron stars, can also have photon spheres. This follows from the fact that the gravitational field of an object does not depend on its actual size, hence any object that is smaller than 1.5 times the Schwarzschild radius corresponding to its mass will indeed have a photon sphere.

Ergosphere

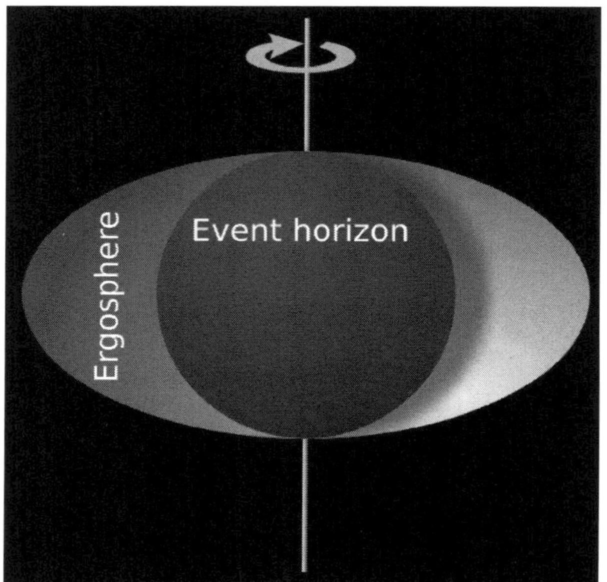

Figure: *The ergosphere is an oblate spheroid region outside of the event horizon, where objects cannot remain stationary.*

Rotating black holes are surrounded by a region of spacetime in which it is impossible to stand still, called the ergosphere. This is the result of a process known as frame-dragging; general relativity predicts that any rotating mass will tend to slightly "drag" along the spacetime immediately surrounding it. Any object near the rotating mass will tend to start moving in the direction of rotation. For a rotating black hole, this effect becomes so strong near the event horizon that an object would have to move faster than the speed of light in the opposite direction to just stand still.

The ergosphere of a black hole is bounded by the (outer) event horizon on the inside and an oblate spheroid, which coincides with the event horizon at the poles and is noticeably wider around the equator. The outer boundary is sometimes called the *ergosurface.*

Objects and radiation can escape normally from the ergosphere. Through the Penrose process, objects can emerge from the ergosphere with more energy than they entered. This energy is taken from the rotational energy of the black hole causing it to slow down.

Formation and Evolution

Considering the exotic nature of black holes, it may be natural to question if such bizarre objects could exist in nature or to suggest that they are merely pathological solutions to Einstein's equations. Einstein himself wrongly thought that black holes would not form, because he held that the angular momentum of collapsing particles would stabilize their motion at some radius. This led the general relativity community to dismiss all results to the contrary for many years. However, a minority of relativists continued to contend that black holes were physical objects, and by the end of the 1960s, they had persuaded the majority of researchers in the field that there is no obstacle to forming an event horizon.

Once an event horizon forms, Penrose proved that a singularity will form somewhere inside it. Shortly afterwards, Hawking showed that many cosmological solutions describing the Big Bang have singularities without scalar fields or other exotic matter. The Kerr solution, the no-hair theorem and the laws of black hole thermodynamics showed that the physical properties of black holes were simple and comprehensible, making them respectable subjects for research. The primary formation process for black holes is expected to be the gravitational collapse of heavy objects such as stars, but there are also more exotic processes that can lead to the production of black holes.

Gravitational Collapse

Gravitational collapse occurs when an object's internal pressure is insufficient to resist the object's own gravity. For stars this usually occurs either because a star has too little "fuel" left to maintain its temperature through stellar nucleosynthesis, or because a star that would have been stable receives extra matter in a way that does not raise its core temperature. In either case the star's temperature is no longer high enough to prevent it from collapsing under its own weight. The collapse may be stopped by the degeneracy pressure of the star's constituents, condensing the matter in an exotic denser state. The result is one of the various types of compact star. The type of compact star formed depends on the mass of the remnant—the matter left over after the outer layers have been blown away, such from a supernova explosion or by pulsations leading to a planetary nebula. Note that this mass can be substantially less than the original star—remnants exceeding 5 solar masses are produced by stars that were over 20 solar masses before the collapse. If the mass of the remnant exceeds about 3–4 solar masses (the Tolman–Oppenheimer–Volkoff limit)—either because the original star was very heavy or because the remnant collected additional mass through accretion of matter—even the degeneracy pressure of neutrons is insufficient to stop the collapse. No known mechanism (except possibly quark degeneracy pressure) is powerful enough to stop the implosion and the object will inevitably collapse to form a black hole.

The gravitational collapse of heavy stars is assumed to be responsible for the formation of stellar mass black holes. Star formation in the early universe may have resulted in very massive stars, which upon their collapse would have produced black holes of up to 10^3 solar masses. These black holes could be the seeds of the supermassive black holes found in the centers of most galaxies.

While most of the energy released during gravitational collapse is emitted very quickly, an outside observer does not actually see the end of this process. Even though the collapse takes a finite amount of time from the reference frame of infalling matter, a distant observer sees the infalling material slow and halt just above the event horizon, due to gravitational time dilation. Light from the collapsing material takes longer and longer to reach the observer, with the light emitted just before the event horizon forms delayed an infinite amount of time. Thus the external observer never sees the formation of the event horizon; instead, the collapsing material seems to become dimmer and increasingly red-shifted, eventually fading away.

Primordial Black Holes in the Big Bang

Gravitational collapse requires great density. In the current epoch of the universe these high densities are only found in stars, but in the early universe shortly after the big bang densities were much greater, possibly allowing for the creation of black holes. The high density alone is not enough to allow the formation of black holes since a uniform mass distribution will not allow the mass to bunch up. In order for primordial black holes to form in such a dense medium, there must be initial density perturbations that can then grow under their own gravity. Different models for the early universe vary widely in their predictions of the size of these perturbations. Various models predict the creation of black holes, ranging from a Planck mass to hundreds of thousands of solar masses. Primordial black holes could thus account for the creation of any type of black hole.

High-Energy Collisions

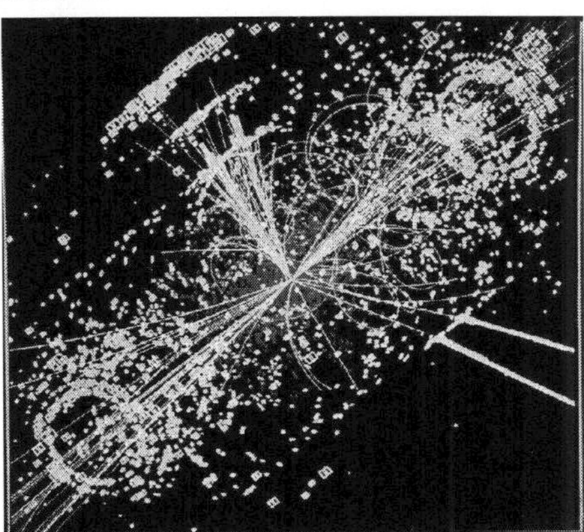

Figure: *A simulated event in the CMS detector, a collision in which a micro black hole may be created.*

Gravitational collapse is not the only process that could create black holes. In principle, black holes could be formed in high-energy collisions that achieve sufficient density. As of 2002, no such events have been detected, either directly or indirectly as a deficiency of the mass balance in particle accelerator experiments. This suggests that there must be a lower limit for the mass of black holes. Theoretically, this boundary is expected to lie around the Planck mass ($m_P = \sqrt{\hbar c/G} \approx 1.2\times10^{19}$ GeV/c^2 $\approx 2.2\times10^{-8}$ kg), where quantum effects are expected to invalidate the predictions of general relativity. This would put the creation of black holes firmly out of reach of any high-energy process occurring on or near the Earth.

However, certain developments in quantum gravity suggest that the Planck mass could be much lower: some braneworld scenarios for example put the boundary as low as 1 TeV/c^2. This would make it conceivable for micro black holes to be created in the high-energy collisions occurring when cosmic rays hit the Earth's atmosphere, or possibly in the new Large Hadron Collider at CERN. Yet these theories are very speculative, and the creation of black holes in these processes is deemed unlikely by many specialists. Even if micro black holes should be formed in these collisions, it is expected that they would evaporate in about 10^{-25} seconds, posing no threat to the Earth.

Growth

Once a black hole has formed, it can continue to grow by absorbing additional matter. Any black hole will continually absorb gas and interstellar dust from its direct surroundings and omnipresent cosmic background radiation. This is the primary process through which supermassive black holes seem to have grown. A similar process has been suggested for the formation of intermediate-mass black holes in globular clusters.

Another possibility is for a black hole to merge with other objects such as stars or even other black holes. Although not necessary for growth, this is thought to have been important, especially for the early development of supermassive black holes, which could have formed from the coagulation of many smaller objects. The process has also been proposed as the origin of some intermediate-mass black holes.

Evaporation

In 1974, Hawking predicted that black holes are not entirely black but emit small amounts of thermal radiation; this effect has become known as Hawking radiation. By applying quantum field theory to a static black hole background, he determined that a black hole should emit particles in a perfect black body spectrum. Since Hawking's publication, many others have verified the result through various approaches. If Hawking's theory of black hole radiation is correct, then black holes are expected to shrink and evaporate over time because they lose mass by the emission of photons and other particles. The temperature of this thermal spectrum (Hawking temperature) is proportional to the surface gravity of the black hole, which, for a Schwarzschild black hole, is inversely proportional to the mass. Hence, large black holes emit less radiation than small black holes.

A stellar black hole of one solar mass has a Hawking temperature of about 100 nanokelvins. This is far less than the 2.7 K temperature of the cosmic microwave background radiation. Stellar-mass or larger black holes receive more mass from the cosmic microwave background than they emit through Hawking radiation and thus will grow instead of shrink. To have a Hawking temperature larger than 2.7 K (and be able to evaporate), a black hole needs to have less mass than the Moon. Such a black hole would have a diameter of less than a tenth of a millimeter.

If a black hole is very small the radiation effects are expected to become very strong. Even a black hole that is heavy compared to a human would evaporate in an instant. A black hole the weight of a car would have a diameter of about 10^{-24} m and take a nanosecond to evaporate, during which time it would briefly have a luminosity more than 200 times that of the Sun. Lower-mass black holes are expected to evaporate even faster; for example, a black hole of mass 1 TeV/c^2

would take less than 10^{-88} seconds to evaporate completely. For such a small black hole, quantum gravitation effects are expected to play an important role and could even—although current developments in quantum gravity do not indicate so—hypothetically make such a small black hole stable.

Observational Evidence

By their very nature, black holes do not directly emit any signals other than the hypothetical Hawking radiation; since the Hawking radiation for an astrophysical black hole is predicted to be very weak, this makes it impossible to directly detect astrophysical black holes from the Earth. A possible exception to the Hawking radiation being weak is the last stage of the evaporation of light (primordial) black holes; searches for such flashes in the past have proven unsuccessful and provide stringent limits on the possibility of existence of light primordial black holes. NASA's Fermi Gamma-ray Space Telescope launched in 2008 will continue the search for these flashes.

Astrophysicists searching for black holes thus have to rely on indirect observations. A black hole's existence can sometimes be inferred by observing its gravitational interactions with its surroundings. A project run by MIT's Haystack Observatory is attempting to observe the event horizon of a black hole directly. Initial results are encouraging.

Accretion of Matter

Figure: *Black hole with corona, X-ray source (artist's concept).*

Due to conservation of angular momentum, gas falling into the gravitational well created by a massive object will typically form a disc-like structure around the object. Friction within the disc causes angular momentum to be transported outward, allowing matter to fall further inward, releasing potential energy and increasing the temperature of the gas.

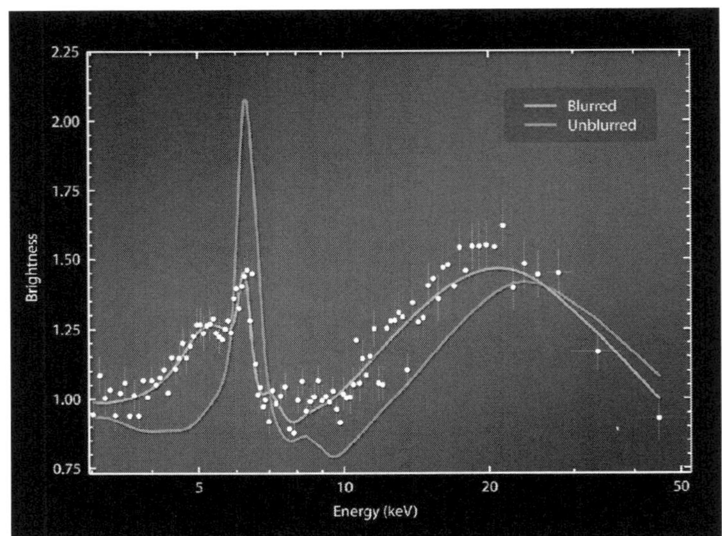

Figure: Blurring of X-rays near Black hole (NuSTAR; 12 August 2014).

In the case of compact objects such as white dwarfs, neutron stars, and black holes, the gas in the inner regions becomes so hot that it will emit vast amounts of radiation (mainly X-rays), which may be detected by telescopes. This process of accretion is one of the most efficient energy-producing processes known; up to 40% of the rest mass of the accreted material can be emitted in radiation. (In nuclear fusion only about 0.7% of the rest mass will be emitted as energy.) In many cases, accretion discs are accompanied by relativistic jets emitted along the poles, which carry away much of the energy. The mechanism for the creation of these jets is currently not well understood.

As such many of the universe's more energetic phenomena have been attributed to the accretion of matter on black holes. In particular, active galactic nuclei and quasars are believed to be the accretion discs of supermassive black holes. Similarly, X-ray binaries are generally accepted to be binary star systems in which one of the two stars is a compact object accreting matter from its companion. It has also been suggested that some ultraluminous X-ray sources may be the accretion disks of intermediate-mass black holes.

X-Ray Binaries

X-ray binaries are binary star systems that are luminous in the X-ray part of the spectrum. These X-ray emissions are generally thought to be caused by one of the component stars being a compact object accreting matter from the other (regular) star. The presence of an ordinary star in such a system provides a unique opportunity for studying the central object and determining if it might be a black hole.

If such a system emits signals that can be directly traced back to the compact object, it cannot be a black hole. The absence of such a signal does, however, not exclude the possibility that the compact object is a neutron star. By studying the companion star it is often possible to obtain the orbital parameters of the system and obtain an estimate for the mass of the compact object. If this is much larger than the Tolman–Oppenheimer–Volkoff limit (that is, the maximum mass a neutron star can have before collapsing) then the object cannot be a neutron star and is generally expected to be a black hole.

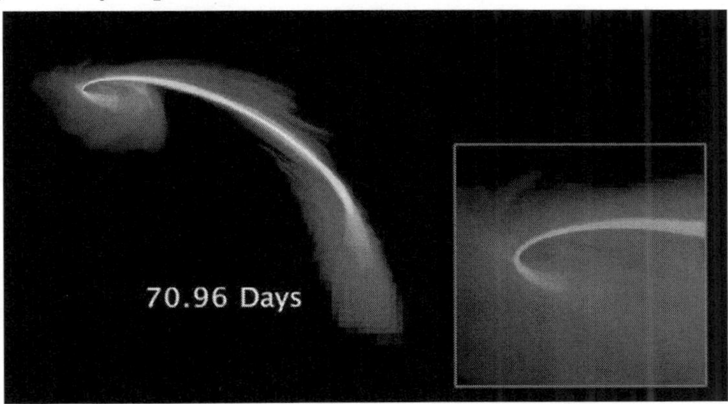

Figure: *A computer simulation of a star being consumed by a black hole. The blue dot indicates the location of the black hole.*

The first strong candidate for a black hole, Cygnus X-1, was discovered in this way by Charles Thomas Bolton, Louise Webster and Paul Murdin in 1972. Some doubt, however, remained due to the uncertainties resultant from the companion star being much heavier than the candidate black hole. Currently, better candidates for black holes are found in a class of X-ray binaries called soft X-ray transients. In this class of system the companion star is relatively low mass allowing for more accurate estimates in the black hole mass. Moreover, these systems are only active in X-ray for several months once every 10–50 years. During the period of low X-ray emission (called quiescence), the accretion disc is extremely faint allowing for detailed observation of the companion star during this period. One of the best such candidates is V404 Cyg.

Quiescence and Advection-Dominated Accretion Flow

The faintness of the accretion disc during quiescence is suspected to be caused by the flow entering a mode called an advection-dominated accretion flow (ADAF). In this mode, almost all the energy generated by friction in the disc is swept along with the flow instead of radiated away. If this model is correct, then it forms strong qualitative evidence for the presence of an event horizon. Because, if the object at the center of the disc had a solid surface, it would emit large amounts of radiation as the highly energetic gas hits the surface, an effect that is observed for neutron stars in a similar state.

Quasi-Periodic Oscillations

The X-ray emission from accretion disks sometimes flickers at certain frequencies. These signals are called quasi-periodic oscillations and are thought to be caused by material moving along the inner edge of the accretion disk (the innermost stable circular orbit). As such their frequency is linked to the mass of the compact object. They can thus be used as an alternative way to determine the mass of potential black holes.

Galactic Nuclei

Astronomers use the term "active galaxy" to describe galaxies with unusual characteristics, such as unusual spectral line emission and very strong radio emission. Theoretical and observational studies have shown that the activity in these active galactic nuclei (AGN) may be explained by the presence of supermassive black holes. The models of these AGN consist of a central black hole that may be millions or billions of times more massive than the Sun; a disk of gas and dust called an accretion disk; and two jets that are perpendicular to the accretion disk.

Although supermassive black holes are expected to be found in most AGN, only some galaxies' nuclei have been more carefully studied in attempts to both identify and measure the actual masses of the central supermassive black hole candidates. Some of the most notable galaxies with supermassive black hole candidates include the Andromeda Galaxy, M32, M87, NGC 3115, NGC 3377, NGC 4258, NGC 4889, NGC 1277, OJ 287, APM 08279+5255 and the Sombrero Galaxy.

It is now widely accepted that the center of nearly every galaxy, not just active ones, contains a supermassive black hole. The close observational correlation between the mass of this hole and the velocity dispersion of the host galaxy's bulge, known as the M-sigma relation, strongly suggests a connection between the formation of the black hole and the galaxy itself.

Currently, the best evidence for a supermassive black hole comes from studying the proper motion of stars near the center of our own

Milky Way. Since 1995 astronomers have tracked the motion of 90 stars in a region called Sagittarius A*. By fitting their motion to Keplerian orbits they were able to infer in 1998 that 2.6 million solar masses must be contained in a volume with a radius of 0.02 lightyears. Since then one of the stars—called S2—has completed a full orbit. From the orbital data they were able to place better constraints on the mass and size of the object causing the orbital motion of stars in the Sagittarius A* region, finding that there is a spherical mass of 4.3 million solar masses contained within a radius of less than 0.002 lightyears. While this is more than 3000 times the Schwarzschild radius corresponding to that mass, it is at least consistent with the central object being a supermassive black hole, and no "realistic cluster [of stars] is physically tenable".

Effects of Strong Gravity

Another way that the black hole nature of an object may be tested in the future is through observation of effects caused by strong gravity in their vicinity. One such effect is gravitational lensing: The deformation of spacetime around a massive object causes light rays to be deflected much like light passing through an optic lens. Observations have been made of weak gravitational lensing, in which light rays are deflected by only a few arcseconds. However, it has never been directly observed for a black hole. One possibility for observing gravitational lensing by a black hole would be to observe stars in orbit around the black hole. There are several candidates for such an observation in orbit around Sagittarius A*.

Another option would be the direct observation of gravitational waves produced by an object falling into a black hole, for example a compact object falling into a supermassive black hole through an extreme mass ratio inspiral. Matching the observed waveform to the predictions of general relativity would allow precision measurements of the mass and angular momentum of the central object, while at the same time testing general relativity. These types of events are a primary target for the proposed Laser Interferometer Space Antenna.

Alternatives

The evidence for stellar black holes strongly relies on the existence of an upper limit for the mass of a neutron star. The size of this limit heavily depends on the assumptions made about the properties of dense matter. New exotic phases of matter could push up this bound. A phase of free quarks at high density might allow the existence of dense quark stars, and some supersymmetric models predict the existence of Q stars.

Some extensions of the standard model posit the existence of preons as fundamental building blocks of quarks and leptons, which could hypothetically form preon stars. These hypothetical models could potentially explain a number of observations of stellar black hole candidates. However, it can be shown from general arguments in general relativity that any such object will have a maximum mass.

Since the average density of a black hole inside its Schwarzschild radius is inversely proportional to the square of its mass, supermassive black holes are much less dense than stellar black holes (the average density of a 10^8 solar mass black hole is comparable to that of water). Consequently, the physics of matter forming a supermassive black hole is much better understood and the possible alternative explanations for supermassive black hole observations are much more mundane. For example, a supermassive black hole could be modelled by a large cluster of very dark objects. However, such alternatives are typically not stable enough to explain the supermassive black hole candidates.

The evidence for stellar and supermassive black holes implies that in order for black holes not to form, general relativity must fail as a theory of gravity, perhaps due to the onset of quantum mechanical corrections. A much anticipated feature of a theory of quantum gravity is that it will not feature singularities or event horizons (and thus no black holes). In 2002, much attention has been drawn by the fuzzball model in string theory. Based on calculations in specific situations in string theory, the proposal suggests that generically the individual states of a black hole solution do not have an event horizon or singularity, but that for a classical/semi-classical observer the statistical average of such states does appear just like an ordinary black hole in general relativity.

Open Questions

Entropy and Thermodynamics

$$S = \frac{1}{4}\frac{c^3 k}{G\hbar}A$$

Figure: *The formula for the Bekenstein–Hawking entropy (S) of a black hole, which depends on the area of the black hole (A). The constants are the speed of light (c), the Boltzmann constant (k), Newton's constant (G), and the reduced Planck constant* $(\hbar)$.

In 1971, Hawking showed under general conditions that the total area of the event horizons of any collection of classical black holes can never decrease, even if they collide and merge. This result, now known as the second law of black hole mechanics, is remarkably similar to the second law of thermodynamics, which states that the total entropy of a system can never decrease. As with classical objects at absolute zero temperature, it was assumed that black holes had zero entropy. If this were the case, the second law of thermodynamics would be violated by entropy-laden matter entering a black hole, resulting in a decrease of the total entropy of the universe. Therefore, Bekenstein proposed that a black hole should have an entropy, and that it should be proportional to its horizon area.

The link with the laws of thermodynamics was further strengthened by Hawking's discovery that quantum field theory predicts that a black hole radiates blackbody radiation at a constant temperature. This seemingly causes a violation of the second law of black hole mechanics, since the radiation will carry away energy from the black hole causing it to shrink. The radiation, however also carries away entropy, and it can be proven under general assumptions that the sum of the entropy of the matter surrounding a black hole and one quarter of the area of the horizon as measured in Planck units is in fact always increasing. This allows the formulation of the first law of black hole mechanics as an analogue of the first law of thermodynamics, with the mass acting as energy, the surface gravity as temperature and the area as entropy.

One puzzling feature is that the entropy of a black hole scales with its area rather than with its volume, since entropy is normally an extensive quantity that scales linearly with the volume of the system. This odd property led Gerard 't Hooft and Leonard Susskind to propose the holographic principle, which suggests that anything that happens in a volume of spacetime can be described by data on the boundary of that volume.

Although general relativity can be used to perform a semi-classical calculation of black hole entropy, this situation is theoretically unsatisfying. In statistical mechanics, entropy is understood as counting the number of microscopic configurations of a system that have the same macroscopic qualities (such as mass, charge, pressure, etc.). Without a satisfactory theory of quantum gravity, one cannot perform such a computation for black holes. Some progress has been made in various approaches to quantum gravity. In 1995, Andrew Strominger and Cumrun Vafa showed that counting the microstates of a specific supersymmetric black hole in string theory reproduced the Bekenstein–Hawking entropy. Since then, similar results have been reported for different black holes both in string theory and in other approaches to quantum gravity like loop quantum gravity.

Information Loss Paradox

Because a black hole has only a few internal parameters, most of the information about the matter that went into forming the black hole is lost. Regardless of the type of matter which goes into a black hole, it appears that only information concerning the total mass, charge, and angular momentum are conserved. As long as black holes were thought to persist forever this information loss is not that problematic, as the information can be thought of as existing inside the black hole, inaccessible from the outside.

However, black holes slowly evaporate by emitting Hawking radiation. This radiation does not appear to carry any additional information about the matter that formed the black hole, meaning that this information appears to be gone forever.

The question whether information is truly lost in black holes (the black hole information paradox) has divided the theoretical physics community. In quantum mechanics, loss of information corresponds to the violation of vital property called unitarity, which has to do with the conservation of probability. It has been argued that loss of unitarity would also imply violation of conservation of energy. Over recent years evidence has been building that indeed information and unitarity are preserved in a full quantum gravitational treatment of the problem.

Magnetometer and Magnetic Measurement Techniques

Magnetometers are measurement instruments used for two general purposes: to measure the magnetization of a magnetic material like a ferromagnet, or to measure the strength and, in some cases, the direction of the magnetic field at a point in space.

The first magnetometer was invented by Carl Friedrich Gauss in 1833 and notable developments in the 19th century included the Hall Effect which is still widely used.

Magnetometers are widely used for measuring the Earth's magnetic field and in geophysical surveys to detect magnetic anomalies of various types. They are also used militarily to detect submarines. Consequently some countries, such as the USA, Canada and Australia classify the more sensitive magnetometers as military technology, and control their distribution.

Magnetometers can be used as metal detectors: they can detect only magnetic (ferrous) metals, but can detect such metals at a much larger depth than conventional metal detectors; they are capable of detecting large objects, such as cars, at tens of metres, while a metal detector's range is rarely more than 2 metres.

In recent years magnetometers have been miniaturized to the extent that they can be incorporated in integrated circuits at very low cost and are finding increasing use as compasses in consumer devices such as mobile phones and tablet computers.

Magnetic fields are vector quantities characterized by both strength and direction. The strength of a magnetic field is measured in units of Tesla in the SI units, and in Gauss in the cgs system of units. 10,000 Gauss are equal to one Tesla. Measurements of the Earth's magnetic field are often quoted in units of nanoTesla (nT), also called a gamma. The Earth's magnetic field can vary from 20,000 to 80,000 nT depending on location, fluctuations in the Earth's magnetic field are on the order of 100 nT, and magnetic field variations due to magnetic anomalies can be in the picoTesla (pT) range. *Gaussmeters* and *teslameters* are magnetometers that measure in units of Gauss or Tesla, respectively. In some contexts, magnetometer is the term used for an instrument that measures fields of less than 1 milliTesla (mT) and gaussmeter is used for those measuring greater than 1 mT.

Types of Magnetometer

There are two basic types of magnetometer measurement. *Vector magnetometers* measure the vector components of a magnetic field. *Total field magnetometers* or *scalar magnetometers* measure the magnitude of the vector magnetic field. Magnetometers used to study the Earth's magnetic field may express the vector components of the field in terms of *declination* (the angle between the horizontal component of the field vector and magnetic north) and the *inclination* (the angle between the field vector and the horizontal surface).

Absolute magnetometers measure the absolute magnitude or vector magnetic field, using an internal calibration or known physical constants of the magnetic sensor. *Relative magnetometers* measure magnitude or vector magnetic field relative to a fixed but uncalibrated baseline. Also called *variometers*, relative magnetometers are used to measure variations in magnetic field.

Magnetometers may also be classified by their situation or intended use. *Stationary magnetometers* are installed to a fixed position and measurements are taken while the magnetometer is stationary.

Portable or *mobile magnetometers* are meant to be used while in motion and may be manually carried or transported in a moving vehicle. *Laboratory magnetometers* are used to measure the magnetic field of materials placed within them and are typically stationary. *Survey magnetometers* are used to measure magnetic fields in geomagnetic surveys; they may be fixed base stations, as in the INTERMAGNET network, or mobile magnetometers used to scan a geographic region.

Performance and Capabilities

The performance and capabilities of magnetometers are described through their technical specifications. Major specifications include

- *Sample rate* is the number of readings given per second. The inverse is the *cycle time* in seconds per reading. Sample rate is important in mobile magnetometers; the sample rate and the vehicle speed determine the distance between measurements.
- *Bandwidth* or *bandpass* characterizes how well a magnetometer tracks rapid changes in magnetic field. For magnetometers with no onboard signal processing, bandwidth is determined by the Nyquist limit set by sample rate. Modern magnetometers may perform smoothing or averaging over sequential samples achieving a lower noise in exchange for lower bandwidth.
- *Resolution* is the smallest change in magnetic field the magnetometer can resolve. A magnetometer should have a resolution a good deal smaller than the smallest change one wishes to observe, to avoid quantization errors.
- *Absolute error* is the difference between the averaged readings of a magnetometer in a constant magnetic field and true magnetic field.
- *Drift* is the change in absolute error over time.
- *Thermal stability* is the dependence of the measurement on temperature. It is given as a temperature coefficient in units of nT per degree Celsius.
- *Noise* is the random fluctuations generated by the magnetometer sensor or electronics. Noise is given in units of $nT/\sqrt{Hz}$, where frequency component refers to the bandwidth.
- *Sensitivity* is the larger of the noise or the resolution.
- *Heading error* is the change in the measurement due to a change in orientation of the instrument in a constant magnetic field.
- The *dead zone* is the angular region of magnetometer orientation in which the instrument produces poor or no measurements.

All optically pumped, proton-free precession, and Overhauser magnetometers experience some dead zone effects.

- *Gradient tolerance* is the ability of a magnetometer to obtain a reliable measurement in the presence of a magnetic field gradient. In surveys of unexploded ordnance or landfills, gradients can be large.

Early Magnetometers

Figure: *Coast and Geodetic Survey Magnetometer No. 18., used in the magnetic survey of Maryland conducted in the 1890s*

In 1833, Carl Friedrich Gauss, head of the Geomagnetic Observatory in Göttingen, published a paper on measurement of the Earth's magnetic field. It described a new instrument that consisted of a permanent bar magnet suspended horizontally from a gold fibre. The difference in the oscillations when the bar was magnetised and when it was demagnetised allowed Gauss to calculate an absolute value for the strength of the Earth's magnetic field.

The gauss, the CGS unit of magnetic flux density was named in his honour, defined as one maxwell per square centimeter; it equals 1×10^{-4} teslas (the SI unit).

Laboratory Magnetometers

Laboratory magnetometers measure the magnetization, also known as the magnetic moment of a sample material. Unlike survey magnetometers, laboratory magnetometers require the sample to be placed inside the magnetometer, and often the temperature, magnetic field, and other parameters of the sample can be controlled. A sample's magnetization, is primarily dependent on the ordering of unpaired electrons within its atoms, with smaller contributions from nuclear magnetic moments, Larmor diamagnetism, among others. Ordering of magnetic moments are primarily classified as diamagnetic, paramagnetic, ferromagnetic, or antiferromagnetic (although the zoology of magnetic ordering also includes ferrimagnetic, helimagnetic, toroidal, spin glasses, etc.). Measuring the magnetization as a function of temperature and magnetic field can give clues as to the type of magnetic ordering, as well as any phase transitions between different types of magnetic orders that occur at critical temperatures or magnetic fields. This type of magnetometry measurement is very important to understand the magnetic properties of materials in physics, chemistry, geophysics and geology, as well as sometimes biology.

SQUID (Superconducting Quantum Interference Device)

SQUIDs are a type of magnetometer used both as survey and as laboratory magnetometers. SQUID magnetometry is an extremely sensitive absolute magnetometry technique. However SQUIDs are noise sensitive, making them impractical as laboratory magnetometers in high DC magnetic fields, and in pulsed magnets. Commercial SQUID magnetometers are available for temperatures between 300 mK and 400 Kelvin, and magnetic fields up to 7 Tesla.

Inductive Pickup Coils

Inductive pickup coils measure the magnetization by detecting the current induced in a coil, due to the changing magnetic moment of the sample. The sample's magnetization can be changed by applying a small applied ac magnetic field, or by a rapidly changing the dc field, as occurs in capacitor-driven pulsed magnets. These measurements require differentiating between the magnetic field produced by the sample and that from the external applied field. Often a special arrangement of cancellation coils is used. For example, half of the pickup coil is wound in one direction, and the other half in the other direction, and the sample is placed in only one half. The external uniform magnetic field will be detected by both halves of the coil and since they are counterwound the external magnetic field produces no net signal.

VSM (Vibrating Sample Magnetometer)

VSM (vibrating sample magnetometers) detect the magnetization of a sample by mechanically vibrating the sample inside of an inductive pickup coil or inside of a SQUID coil. Induced current or changing flux in the coil is measured. The vibration is typically created by a motor or a piezoelectric actuator. Typically the VSM technique is about an order of magnitude less sensitive than SQUID magnetometry. VSMs can be combined with SQUIDs to create a system that is more sensitive than either one alone. Heat due to the sample vibration can limit the base temperature of a VSM, typically to 2 Kelvin. VSM is also impractical for measuring a fragile sample that is sensitive to rapid acceleration.

Pulsed Field Extraction Magnetometry

Pulsed Field Extraction Magnetometry is another method making use of pickup coils to measure magnetization. Unlike VSMs where the sample is physically vibrated, in Pulsed Field Extraction Magnetometry, the sample is secured and the external magnetic field is changed rapidly, for example in a capacitor-driven magnet. One of multiple techniques must then be used to cancel out the external field from the field produced by the sample. These include counterwound coils that cancel the external uniform field, and background measurements with the sample removed from the coil.

Torque Magnetometry

Magnetic torque magnetometry can be even more sensitive than SQUID magnetometry. However, magnetic torque magnetometry doesn't measure magnetism directly as all the previously mentioned methods do. Magnetic torque magnetometry instead measures the torque τ acting on a sample's magnetic moment μ as a result of a uniform magnetic field B, $\tau=\mu\times B$. A torque is thus a measure of the sample's magnetic or shape anisotropy. In some cases the sample's magnetization can be extracted from the measured torque. In other cases, the magnetic torque measurement is used to detected magnetic phase transitions or quantum oscillations. The most common way to measure magnetic torque is to mount the sample on a cantilever and measure the displacement via capacitance measurement between the cantilever and nearby fixed object, or by measuring the piezoelectricity of the cantilever, or by optical interferometry off the surface of the cantilever.

Faraday Force Magnetometry

Faraday Force Magnetometry makes use of the fact that a spatial magnetic field gradient will produce force acting on a magnetized object,

$F = (M \cdot \nabla) B$. In Faraday Force Magnetometry the force on the sample can be measured by a scale (hanging the sample from a sensitive balance), or by detecting the displacement against a spring. Commonly a capacitive load cell or cantilever is used because of its sensitivity, size, and lack of mechanical parts. Faraday Force Magnetometry is approximately one order of magnitude less sensitive than a SQUID. The biggest drawback to Faraday Force Magnetometry is that it requires some means of not only producing a magnetic field, but also producing a magnetic field gradient. While this can be accomplished by using a set of special pole faces, a much better result can be achieved by using set of gradient coils. A major advantage to Faraday Force Magnetometry is that it is small and reasonably tolerant to noise, and thus can be implemented in a wide range of environments, including a dilution refrigerator. Faraday Force Magnetometry can also be complicated by the presence of torque. This can be circumvented by varying the gradient field independently of the applied DC field so the torque and the Faraday Force contribution can be separated, and/or by designing a Faraday Force Magnetometer that prevents the sample from being rotated.

Optical Magnetometry

Optical magnetometry makes use of various optical techniques to measure magnetization. One such technique, Kerr Magnetometry makes use of the magneto-optic Kerr effect, or MOKE. In this technique, incident light is directed at the sample's surface. Light interacts with a magnetized surface nonlinearly so the reflected light has an elliptical polarization which is then measured by a detector. Another method of optical magnetometry is Faraday Rotation Magnetometry. Faraday Rotation Magnetometry utilises nonlinear magneto-optical rotation to measure a sample's magnetization. In this method a Faraday Modulating thin film is applied to the sample to be measured and a series of images are taken with a camera that senses the polarization of the reflected light. In order to reduce noise, multiple pictures are then averaged together. One advantage to this method is that it allows mapping of the magnetic characteristics over the surface of a sample. This can be especially useful when studying such things as the Meissner Effect on superconductors.

Survey Magnetometers

Survey magnetometers can be divided into two basic types:

- *Scalar magnetometers* measure the total strength of the magnetic field to which they are subjected, but not its direction

- *Vector magnetometers* have the capability to measure the component of the magnetic field in a particular direction, relative to the spatial orientation of the device.

A vector is a mathematical entity with both magnitude and direction. The Earth's magnetic field at a given point is a vector. A magnetic compass is designed to give a horizontal bearing direction, whereas a *vector magnetometer* measures both the magnitude and direction of the total magnetic field. Three orthogonal sensors are required to measure the components of the magnetic field in all three dimensions.

They are also rated as "absolute" if the strength of the field can be calibrated from their own known internal constants or "relative" if they need to be calibrated by reference to a known field.

A *magnetograph* is a magnetometer that continuously records data. Magnetometers can also be classified as "AC" if they measure fields that vary relatively rapidly in time (>100 Hz), and "DC" if they measure fields that vary only slowly (quasi-static) or are static. AC magnetometers find use in electromagnetic systems (such as magnetotellurics), and DC magnetometers are used for detecting mineralisation and corresponding geological structures.

Scalar Magnetometers

Proton Precession Magnetometer

Proton precession magnetometers, also known as *proton magnetometers*, PPMs or simply mags, measure the resonance frequency of protons (hydrogen nuclei) in the magnetic field to be measured, due to nuclear magnetic resonance (NMR). Because the precession frequency depends only on atomic constants and the strength of the ambient magnetic field, the accuracy of this type of magnetometer can reach 1 ppm.

A direct current flowing in a solenoid creates a strong magnetic field around a hydrogen-rich fluid (kerosene and decane are popular, and even water can be used), causing some of the protons to align themselves with that field. The current is then interrupted, and as protons realign themselves with the ambient magnetic field, they precess at a frequency that is directly proportional to the magnetic field. This produces a weak rotating magnetic field that is picked up by a (sometimes separate) inductor, amplified electronically, and fed to a digital frequency counter whose output is typically scaled and displayed directly as field strength or output as digital data.

For hand/backpack carried units, PPM sample rates are typically limited to less than one sample per second. Measurements are typically taken with the sensor held at fixed locations at approximately 10 metre increments.

Portable instruments are also limited by sensor volume (weight) and power consumption. PPMs work in field gradients up to 3,000 nT/m which is adequate from most mineral exploration work. For higher gradient tolerance, such as mapping banded iron formations and detecting large ferrous objects, Overhauser magnetometers can handle 10,000 nT/m, and caesium magnetometers can handle 30,000 nT/m.

They are relatively inexpensive (< 8,000 USD) and were once widely used in mineral exploration. Three manufacturers dominate the market: GEM Systems, Geometrics and Scintrex. Popular models include G-856, Smartmag and GSM-18 and GSM-19T.

For mineral exploration, they have been superseded by Overhauser, Caesium and Potassium instruments, all of which are fast-cycling, and do not require the operator to pause between readings.

Overhauser Effect Magnetometer

The *Overhauser effect magnetometer* or *Overhauser magnetometer* uses the same fundamental effect as the *proton precession magnetometer* to take measurements. By adding free radicals to the measurement fluid, the nuclear Overhauser effect can be exploited to significantly improve upon the proton precession magnetometer. Rather than aligning the protons using a solenoid, a low power radio-frequency field is used to align (polarise) the electron spin of the free radicals, which then couples to the protons via the Overhauser effect. This has two main advantages: driving the RF field takes a fraction of the energy (allowing lighter-weight batteries for portable units), and faster sampling as the electron-proton coupling can happen even as measurements are being taken. An Overhauser magnetometer produces readings with a 0.01 nT to 0.02 nT standard deviation while sampling once per second.

Caesium Vapour Magnetometer

The *optically pumped caesium vapour magnetometer* is a highly sensitive (300 fT/Hz$^{0.5}$) and accurate device used in a wide range of applications. It is one of a number of alkali vapours (including rubidium and potassium) that are used in this way, as well as helium.

The device broadly consists of a photon emitter containing a caesium light emitter or lamp, an absorption chamber containing caesium vapour, a "buffer gas" through which the emitted photons pass and a photon detector, arranged in that order.

The basic principle that allows the device to operate is the fact that a caesium atom can exist in any of nine energy levels, which can be informally thought of as the placement of electron atomic orbitals around the atomic nucleus. When a caesium atom within the chamber encounters a photon from the lamp, it is excited to a higher energy state, emits a photon and falls to an indeterminate lower energy state. The caesium atom is "sensitive" to the photons from the lamp in three of its nine energy states, and therefore, assuming a closed system, all the atoms will eventually fall into a state in which all the photons from the lamp will pass through unhindered and be measured by the photon detector. At this point, the sample (or population) is said to be polarized and ready for measurement to take place. This process is done continuously during operation. This theoretically perfect magnetometer is now functional and so can begin to make measurements.

In the most common type of caesium magnetometer, a very small AC magnetic field is applied to the cell. Since the difference in the energy levels of the electrons is determined by the external magnetic field, there is a frequency at which this small AC field will cause the electrons to change states. In this new state, the electron will once again be able to absorb a photon of light. This causes a signal on a photo detector that measures the light passing through the cell. The associated electronics use this fact to create a signal exactly at the frequency which corresponds to the external field.

Another type of caesium magnetometer modulates the light applied to the cell. This is referred to as a Bell-Bloom magnetometer, after the two scientists who first investigated the effect. If the light is turned on and off at the frequency corresponding to the Earth's field, there is a change in the signal seen at the photo detector. Again, the associated electronics use this to create a signal exactly at the frequency which corresponds to the external field. Both methods lead to high performance magnetometers.

Potassium Vapour Magnetometer

Potassium is the only optically pumped magnetometer that operates on a single, narrow electron spin resonance (ESR) line is contrast to other alkali vapour magnetometers that use an irregular, composite and wide spectral lines and Helium with the inherently wide spectral line.

Applications

The caesium and potassium magnetometers are typically used where a higher performance magnetometer than the proton magnetometer is needed. In archaeology and geophysics, where the sensor sweeps through an area and many accurate magnetic field measurements are often needed, caesium and potassium magnetometers have advantages over the proton magnetometer.

The caesium and potassium magnetometer's faster measurement rate allows the sensor to be moved through the area more quickly for a given number of data points. Caesium and potassium magnetometers are insensitive to rotation of the sensor while the measurement is being made. The lower noise of caesium and potassium magnetometers allow those measurements to more accurately show the variations in the field with position.

Vector Magnetometers

Vector magnetometers measure one or more components of the magnetic field electronically. Using three orthogonal magnetometers, both azimuth and dip (inclination) can be measured. By taking the square root of the sum of the squares of the components the total magnetic field strength (also called total magnetic intensity, TMI) can be calculated by Pythagoras's theorem.

Vector magnetometers are subject to temperature drift and the dimensional instability of the ferrite cores. They also require levelling to obtain component information, unlike total field (scalar) instruments. For these reasons they are no longer used for mineral exploration.

Rotating Coil Magnetometer

The magnetic field induces a sine wave in a rotating coil. The amplitude of the signal is proportional to the strength of the field, provided it is uniform, and to the sine of the angle between the rotation axis of the coil and the field lines. This type of magnetometer is obsolete.

Hall Effect Magnetometer

Main Article: Hall E\ffect Sensor

The most common magnetic sensing devices are solid-state Hall effect sensors. These sensors produce a voltage proportional to the applied magnetic field and also sense polarity. They are used in applications where the magnetic field strength is relatively large, such as in anti-lock braking systems in cars which sense wheel rotation speed via slots in the wheel disks.

Magnetoresistive Devices

These are made of thin strips of permalloy (NiFe magnetic film) whose electrical resistance varies with a change in magnetic field. They have a well-defined axis of sensitivity, can be produced in 3-D versions and can be mass-produced as an integrated circuit. They have a response time of less than 1 microsecond and can be sampled in moving vehicles up to 1,000 times/second. They can be used in compasses that read within 1°, for which the underlying sensor must reliably resolve 0.1°.

Fluxgate Magnetometer

Fluxgate magnetometers were invented in the 1930s by Victor Vacquier at Gulf Research Laboratories. Vacquier applied them during World War II as an instrument for detecting submarines, and after the war confirmed the theory of plate tectonics by using them to measure shifts in the magnetic patterns on the sea floor.

Figure: *A uniaxial fluxgate magnetometer.*

A fluxgate magnetometer consists of a small, magnetically susceptible core wrapped by two coils of wire. An alternating electrical current is passed through one coil, driving the core through an alternating cycle of magnetic saturation; i.e., magnetised, unmagnetised, inversely magnetised, unmagnetised, magnetised, and so forth. This constantly changing field induces an electrical current in the second coil, and this output current is measured by a detector. In a magnetically neutral background, the input and output currents will match. However, when the core is exposed to a background field, it will be more easily saturated in alignment with that field and less easily saturated in opposition to it. Hence the alternating magnetic field, and the induced output current, will be out of step with the input current. The extent to which this is the case will depend on the strength of the background magnetic field. Often, the current in the output coil is integrated, yielding an output analog voltage, proportional to the magnetic field.

Figure: *A fluxgate compass/inclinometer.*

A wide variety of sensors are currently available and used to measure magnetic fields. Fluxgate compasses and gradiometers measure the direction and magnitude of magnetic fields. Fluxgates are affordable, rugged and compact. This, plus their typically low power consumption makes them ideal for a variety of sensing applications. Gradiometers are commonly used for archaeological prospecting and unexploded ordinance (UXO) detection such as the German military's popular *Foerster*.

The typical fluxgate magnetometer consists of a "sense" (secondary) coil surrounding an inner "drive" (primary) coil that is wound around permeable core material. Each sensor has magnetic core elements that can be viewed as two carefully matched halves. An alternating current is applied to the drive winding, which drives the core into plus and minus saturation. The instantaneous drive current in each core half is driven in opposite polarity with respect to any external magnetic field. In the absence of any external magnetic field, the flux in one core half cancels that in the other, and so the total flux seen by the sense coil is zero. If an external magnetic field is now applied, it will, at a given instance in time, aid the flux in one core half and oppose flux in the other. This causes a net flux imbalance between the halves, so that they no longer cancel one another. Current pulses are now induced in the sense coil winding on every drive current phase reversal (or at the 2nd, and all even harmonics). This results in a signal that is dependent on both the external field magnitude and polarity.

There are additional factors that affect the size of the resultant signal. These factors include the number of turns in the sense winding, magnetic permeability of the core, sensor geometry and the gated flux rate of change with respect to time. Phase synchronous detection is used to convert these harmonic signals to a DC voltage proportional to the external magnetic field.

SQUID

A SQUID (for superconducting quantum interference device) is a very sensitive magnetometer used to measure extremely subtle magnetic fields, based on superconducting loops containing Josephson junctions.

SQUIDs are sensitive enough to measure fields as low as 5 aT (5×10^{-18} T) within a few days of averaged measurements. Their noise levels are as low as 3 fT$\cdot$Hz$^{-½}$. For comparison, a typical refrigerator magnet produces 0.01 teslas (10^{-2} T), and some processes in animals produce very small magnetic fields between 10^{-9} T and 10^{-6} T.

Recently invented SERF atomic magnetometers are potentially more sensitive and do not require cryogenic refrigeration but are orders of magnitude larger in size (~1 cm^3) and must be operated in a near-zero magnetic field.

SQUID Magnetometer

SQUIDs, or superconducting quantum interference devices, measure extremely small changes in magnetic fields. They are very sensitive vector magnetometers, with noise levels as low as 3 fT Hz$^{-½}$ in commercial instruments and 0.4 fT Hz$^{-½}$ in experimental devices.

Many liquid-helium-cooled commercial SQUIDs achieve a flat noise spectrum from near DC (less than 1 Hz) to tens of kilohertz, making such devices ideal for time-domain biomagnetic signal measurements. SERF atomic magnetometers demonstrated in laboratories so far reach competitive noise floor but in relatively small frequency ranges.

SQUID magnetometers require cooling with liquid helium (4.2 K) or liquid nitrogen (77 K) to operate, hence the packaging requirements to use them are rather stringent both from a thermal-mechanical as well as magnetic standpoint. SQUID magnetometers are most commonly used to measure the magnetic fields produced by laboratory samples, also for brain or heart activity (magnetoencephalography and magnetocardiography, respectively).

Geophysical surveys use SQUIDS from time to time, but the logistics of cooling the SQUID are much more complicated than other magnetometers that operate at room temperature.

Spin-Exchange Relaxation-Free (SERF) Atomic Magnetometers

At sufficiently high atomic density, extremely high sensitivity can be achieved. Spin-exchange-relaxation-free (SERF) atomic magnetometers containing potassium, caesium or rubidium vapor operate similarly to the caesium magnetometers described above, yet can reach sensitivities lower than 1 fT $Hz^{-½}$. The SERF magnetometers only operate in small magnetic fields. The Earth's field is about 50 μT; SERF magnetometers operate in fields less than 0.5 μT.

Large volume detectors have achieved a sensitivity of 200 aT $Hz^{-½}$. This technology has greater sensitivity per unit volume than SQUID detectors. The technology can also produce very small magnetometers that may in the future replace coils for detecting changing magnetic fields. This technology may produce a magnetic sensor that has all of its input and output signals in the form of light on fiber-optic cables. This would allow the magnetic measurement to be made in places where high electrical voltages exist.

Uses

Magnetometers have a very diverse range of applications, including locating objects such as submarines, sunken ships, hazards for tunnel boring machines, hazards in coal mines, unexploded ordnance, toxic waste drums, as well as a wide range of mineral deposits and geological structures. They also have applications in heart beat monitors, weapon systems positioning, sensors in anti-locking brakes, weather prediction (via solar cycles), steel pylons, drill guidance systems, archaeology, plate tectonics and radio wave propagation and planetary exploration.

Depending on the application, magnetometers can be deployed in spacecraft, aeroplanes (*fixed wing* magnetometers), helicopters (*stinger* and *bird*), on the ground (*backpack*), towed at a distance behind quad bikes (*sled* or *trailer*), lowered into boreholes (*tool*, *probe* or *sonde*) and towed behind boats (*tow fish*).

Archaeology

Magnetometers are also used to detect archaeological sites, shipwrecks and other buried or submerged objects. Fluxgate gradiometers are popular due to their compact configuration and relatively low cost. Gradiometers enhance shallow features and negate the need for a base station.

Caesium and Overhauser magnetometers are also very effective when used as gradiometers or as single-sensor systems with base stations.

The TV program *Time Team* popularised 'geophys', including magnetic techniques used in archaeological work to detect fire hearths, walls of baked bricks and magnetic stones such as basalt and granite. Walking tracks and roadways can sometimes be mapped with differential compaction in magnetic soils or with disturbances in clays, such as on the Great Hungarian Plain. Ploughed fields behave as sources of magnetic noise in such surveys.

Auroras

Magnetometers can give an indication of auroral activity before the light from the aurora becomes visible. A grid of magnetometers around the world constantly measures the effect of the solar wind on the Earth's magnetic field, which is then published on the K-index.

Coal Exploration

Whilst magnetometers can be used to help map basin shape at a regional scale, they are more commonly used to map hazards to coal mining, such as basaltic intrusions (dykes, sills and volcanic plugs) that destroy resources and are dangerous to longwall mining equipment. Magnetometers can also locate zones ignited by lightning and map siderite (an impurity in coal).

The best survey results are achieved on the ground in high-resolution surveys (with approximately 10 m line spacing and 0.5 m station spacing). Bore-hole magnetometers using a Ferret can also assist when coal seams are deep, by using multiple sills or looking beneath surface basalt flows.

Modern surveys generally use magnetometers with GPS technology to automatically record the magnetic field and their location. The data set is then corrected with data from a second magnetometer (the base station) that is left stationary and records the change in the Earth's magnetic field during the survey.

Directional Drilling

Magnetometers are used in directional drilling for oil or gas to detect the azimuth of the drilling tools near the drill. They are most often paired with accelerometers in drilling tools so that both the inclination and azimuth of the drill can be found.

Military

For defencive purposes, navies use arrays of magnetometers laid across sea floors in strategic locations (i.e. around ports) to monitor submarine activity. The Russian 'Goldfish' (titanium submarines) were designed and built at great expense to thwart such systems (as pure titanium is non-magnetic).

Military submarines are degaussed by passing through large underwater loops at regular intervals in a bid in order to escape detection by sea-floor monitoring systems, magnetic anomaly detectors and mines that are triggered by magnetic anomalies. However, submarines are never completely de-magnetised. It is possible to tell the depth at which a submarine has been by measuring its magnetic field, which is distorted as the pressure distorts the hull and hence the field. Heating can also change the magnetization of steel.

Submarines tow long sonar arrays to detect ships, and can even recognise different propeller noises. The sonar arrays need to be accurately positioned so they can triangulate direction to targets (e.g. ships). The arrays do not tow in a straight line, so fluxgate magnetometers are used to orient each sonar node in the array.

Fluxgates can also be used in weapons navigation systems, but have been largely superseded by GPS and ring laser gyroscopes.

Magnetometers such as the German Foerster are used to locate ferrous ordnance. Caesium and Overhauser magnetometers are used to locate and help clean up old bombing/test ranges.

UAV payloads also include magnetometers for a range of defencive and offencive tasks.

Mineral Exploration

Magnetometric surveys can be useful in defining magnetic anomalies which represent ore (direct detection), or in some cases gangue minerals associated with ore deposits (indirect or inferential detection). This includes iron ore, magnetite, hematite and often pyrrhotite.

First world countries such as Australia, Canada and USA invest heavily in systematic airborne magnetic surveys of their respective continents and surrounding oceans, to assist with map geology and in the discovery of mineral deposits. Such aeromag surveys are typically undertaken with 400 m line spacing at 100 m elevation, with readings every 10 meters or more. To overcome the asymmetry in the data density, data is interpolated between lines (usually 5 times) and data along the line is then averaged. Such data would be gridded to an 80 m × 80 m pixel size and image processed using a program like ERMapper. At an exploration lease scale, the survey may be followed by a more detailed helimag or crop duster style fixed wing at 50 m line spacing and 50 m elevation (terrain permitting). Such an image would be gridded on a 10 x 10 m pixel, offering 64 times the resolution.

Where targets are shallow (<200 m), aeromag anomalies may be followed up with ground magnetic surveys on 10 m to 50 m line spacing with 1 m station spacing in order to give the best detail (2 to 10 m pixel grid) (or 25 times the resolution prior to drilling).

Magnetic fields from magnetic bodies of ore fall off with the inverse distance cubed (dipole target), or at best inverse distance squared (magnetic monopole target). One analogy to the resolution-with-distance is a car driving at night with lights on. At a distance of 400 m one sees one glowing haze, but as it approaches, two headlights, and then the left blinker, are visible.

There are many challenges interpreting magnetic data for mineral exploration. Multiple targets mix together like multiple heat sources and, unlike light, there is no magnetic telescope to focus fields. The combination of multiple sources is measured at the surface. The geometry, depth or magnetisation direction (remanence) of the targets are also generally not known, and so multiple models can explain the data.

Potent by Geophysical Software Solutions [1] is a leading magnetic (and gravity) interpretation package used extensively in the Australian exploration industry.

Magnetometers assist mineral explorers both directly (i.e., gold mineralisation associated with magnetite, diamonds in kimberlite pipes) and, more commonly, indirectly, such as by mapping geological structures conducive to mineralisation (i.e., shear zones and alteration haloes around granites).

Airborne Magnetometers detect the change in the Earth's magnetic field using sensors attached to the aircraft in the form of a "stinger" or by towing a magnetometer on the end of a cable. The magnetometer on a cable is often referred to as a "bomb" because of its shape. Others call it a "bird".

Because hills and valleys under the aircraft will cause the magnetic readings to rise and fall, a radar altimeter is used to keep track of the transducer's deviation from the nominal altitude above ground. There may also be a camera that takes photos of the ground. The location of the measurement is determined by also recording a GPS.

Mobile Telephones

Many smartphones contain magnetometers; apps exist that serve as compasses. The iPhone 3GS has a magnetometer, a magnetoresistive permalloy sensor, the AN-203 produced by Honeywell. In 2009, the price of three-axis magnetometers dipped below US $1 per device and dropped rapidly.

The use of a three-axis device means that it is not sensitive to the way it is held in orientation or elevation. Hall effect devices are also popular.

Researchers at Deutsche Telekom have used magnetometers embedded in mobile devices to permit touchless 3D interaction. Their interaction framework, called MagiTact, tracks changes to the magnetic field around a cellphone to identify different gestures made by a hand holding or wearing a magnet.

Oil Exploration

Seismic methods are preferred to magnetometers as the primary survey method for oil exploration although magnetic methods can give additional information about the underlying geology and in some environments evidence of leakage from traps. Magnetometers are also used in oil exploration to show locations of geologic features that would make drilling impractical, and other features that give geophysicists a more complete picture of stratigraphy.

Spacecraft

A three-axis fluxgate magnetometer was part of the Mariner 2 and Mariner 10 missions. A dual technique magnetometer is part of the Cassini–Huygens mission to explore Saturn.

This system is composed of a vector helium and fluxgate magnetometers. Magnetometers are also a component instrument on the Mercury MESSENGER mission. A magnetometer can also be used by satellites like GOES to measure both the magnitude and direction of the magnetic field of a planet or moon.

Magnetic Surveys

Systematic surveys can be used to in searching for mineral deposits or locating lost objects. Such surveys are divided into:

- Aeromagnetic survey
- Borehole
- Ground
- Marine

Data can be divided in point located and image data, the latter of which is in ERMapper format.

Magnetovision

On the base of space measured distribution of magnetic field parameters (e.g. amplitude or direction), the magnetovision images may be generated. Such presentation of magnetic data is very useful for further analyse and data fusion.

Gradiometer

Magnetic gradiometers are pairs of magnetometers with their sensors separated, usually horizontally, by a fixed distance. The readings are subtracted in order to measure the difference between the sensed magnetic fields, which gives the field gradients caused by magnetic anomalies.

This is one way of compensating both for the variability in time of the Earth's magnetic field and for other sources of electromagnetic interference, thus allowing for more sensitive detection of anomalies. Because nearly equal values are being subtracted, the noise performance requirements for the magnetometers is more extreme.

Gradiometers enhance shallow magnetic anomalies and are thus good for archaeological and site investigation work. They are also good for real-time work such as unexploded ordnance location. It is twice as efficient to run a base station and use two (or more) mobile sensors to read parallel lines simultaneously (assuming data is stored and post-processed).

In this manner, both along-line and cross-line gradients can be calculated.

Position Control of Magnetic Surveys

In traditional mineral exploration and archaeological work, grid pegs placed by theodolite and tape measure were used to define the survey area. Some UXO surveys used ropes to define the lanes. Airborne surveys used radio triangulation beacons, such as Siledus.

Non-magnetic electronic hipchain triggers were developed to trigger magnetometers. They used rotary shaft encoders to measure distance along disposable cotton reels.

Modern explorers use a range of low-magnetic signature GPS units, including Real-Time Kinematic GPS.

Heading Errors in Magnetic Surveys

Magnetic surveys can suffer from noise coming from a range of sources. Different magnetometer technologies suffer different kinds of noise problems.

Heading errors are one group of noise. They can come from three sources:

- Sensor
- Console
- perator

Some total field sensors give different readings depending on their orientation. Magnetic materials in the sensor itself are the primary cause of this error. In some magnetometers, such as the vapor magnetometers (caesium, potassium, etc.), there are sources of heading error in the physics that contribute small amounts to the total heading error.

Console noise comes from magnetic components on or within the console. These include ferrite in cores in inductors and transformers, steel frames around LCD's, legs on IC chips and steel cases in disposable batteries. Some popular MIL spec connectors also have steel springs.

Operators must take care to be magnetically clean and should check the 'magnetic hygiene' of all apparel and items carries during a survey. Akubra hats are very popular in Australia, but their steel rims must be removed before use on magnetic surveys. Steel rings on notepads, steel capped boots and steel springs in overall eyelets can all cause unnecessary noise in surveys. Pens, mobile phones and stainless steel implants can also be problematic.

The magnetic response (noise) from ferrous object on the operator and console can change with heading direction because of induction and remanence. Aeromagnetic survey aircraft and quad bike systems can use special compensators to correct for heading error noise.

Heading errors look like herringbone patterns in survey images. Alternate lines can also be corrugated.

Image Processing of Magnetic Data

Recording data and image processing is superior to real-time work because subtle anomalies often missed by the operator (especially in magnetically noisy areas) can be correlated between lines, shapes and clusters better defined. A range of sophisticated enhancement techniques can also be used. There is also a hard copy and need for systematic coverage.

Earth's Field NMR

Nuclear magnetic resonance (NMR) in the geomagnetic field is conventionally referred to as Earth's field NMR (EFNMR). EFNMR is a special case of low field NMR.

When a sample is placed in a constant magnetic field and stimulated (perturbed) by a pulsed or alternating magnetic field, NMR active nuclei resonate at characteristic frequencies. Examples of such nuclei are the isotopes carbon-13, and hydrogen-1 also referred to as protons. The resonant frequency of each isotope is directly proportional to the strength of the applied magnetic field, and the magnetogyric or gyromagnetic ratio of that isotope. The signal strength is proportional both to the stimulating magnetic field and the number of nuclei of that isotope in the sample. Thus in the 21 tesla magnetic field that may be found in high resolution laboratory NMR spectrometers, protons resonate at 900 MHz. However in the Earth's magnetic field the same nuclei resonate at audio frequencies of around 2 kHz and generate very weak signals.

The location of a nucleus within a complex molecule affects the 'chemical environment' (i.e. the rotating magnetic fields generated by the other nuclei) experienced by the nucleus. Thus different hydrocarbon molecules containing NMR active nuclei in different positions within the molecules produce slightly different patterns of resonant frequencies.

EFNMR signals can be affected by both magnetically noisy laboratory environments and natural variations in the Earth's field, which originally compromised its usefulness. However this disadvantage has been overcome by the introduction of electronic equipment which compensates changes in ambient magnetic fields.

Whereas chemical shifts are important in NMR, they are insignificant in the Earth's field. The absence of chemical shifts causes features such as spin-spin multiplets (that are separated by high fields) to be superimposed in EFNMR. Instead, EFNMR spectra are dominated by spin-spin coupling (J-coupling) effects. Software optimised for analysing these spectra can provide useful information about the structure of the molecules in the sample.

Applications

Applications of EFNMR include:

- Proton precession magnetometers (PPM) or proton magnetometers, which produce magnetic resonance in a known sample in the magnetic field to be measured, measure the sample's resonant frequency, then calculate and display the field strength.
- EFNMR spectrometers, which use the principle of NMR spectroscopy to analyse molecular structures in a variety of

applications, from investigating the structure of ice crystals in polar ice-fields, to rocks and hydrocarbons on-site.

- Earth's field MRI scanners, which use the principle of magnetic resonance imaging.

The advantages of the Earth's field instruments over conventional (high field strength) instruments include the portability of the equipment giving the ability to analyse substances on-site, and their lower cost. The much lower geomagnetic field strength, that would otherwise result in poor signal-to-noise ratios, is compensated by homogeneity of the Earth's field giving the ability to use much larger samples. Their relatively low cost and simplicity make them good educational tools.

Examples (illustrated) are the TeachSpin and Magritek's Terranova-MRI instruments.

Although those commercial EFNMR spectrometers and MRI instruments aimed at universities etc. are necessarily sophisticated and are too costly for most hobbyists, internet search engines find data and designs for basic proton precession magnetometers which claim to be within the capability of reasonably competent electronic hobbyists or undergraduate students to build from readily available components costing no more than a few tens of US dollars.

Mode of Operation

Free Induction Decay (FID) is the magnetic resonance due to Larmor precession that results from the stimulation of nuclei by means of either a *pulsed dc magnetic field* or a *pulsed resonant frequency (rf) magnetic field*, somewhat analogous respectively to the effects of plucking or bowing a stringed instrument. Whereas a pulsed rf field is usual in conventional (high field) NMR spectrometers, the pulsed dc polarising field method of stimulating FID is usual in EFNMR spectrometers and PPMs.

EFNMR equipment typically incorporates several coils, for stimulating the samples and for sensing the resulting NMR signals. Signal levels are very low, and specialised electronic amplifiers are required to amplify the EFNMR signals to usable levels. The stronger the polarising magnetic field, the stronger the EFNMR signals and the better the signal-to-noise ratios. The main trade-offs are performance versus portability and cost.

Since the FID resonant frequencies of NMR active nuclei are directly proportional to the magnetic field affecting those nuclei, we can use widely available NMR spectroscopy data to analyse suitable substances in the Earth's magnetic field.

An important feature of EFNMR compared with high-field NMR is that some aspects of molecular structure can be observed more clearly at low fields and low frequencies, whereas other features observable at high fields may not be observable at low fields. This is because:

- Electron-mediated heteronuclear J-couplings (spin-spin couplings) are field independent, producing clusters of two or more frequencies separated by several Hz, which are more easily observed in a fundamental resonance of about 2 kHz. "Indeed it appears that enhanced resolution is possible due to the long spin relaxation times and high field homogeneity which prevail in EFNMR."
- Chemical shifts of several parts per million (ppm) are clearly separated in high field NMR spectra, but have separations of only a few milliherz at proton EFNMR frequencies, and so are undetectable in an experiment that takes place on a timescale of tenths of a second.

Proton EFNMR Frequencies

The geomagnetic field strength and hence precession frequency varies with location and time.

Larmor precession frequency = magnetogyric ratio x magnetic field

Proton magnetogyric ratio = 42.576 Hz/μT (also written 42.576 MHz/T or 0.042576 Hz/nT)

Earth's magnetic field: 30 μT near Equator to 60 μT near Poles, around 50 μT at mid-latitudes.

Thus proton (hydrogen nucleus) EFNMR frequencies are audio frequencies of about 1.3 kHz near the Equator to 2.5 kHz near the Poles, around 2 kHz being typical of mid-latitudes. In terms of the electromagnetic spectrum EFNMR frequencies are in the VLF and ULF radio frequency bands.

Examples of molecules containing hydrogen nuclei useful in proton EFNMR are water, hydrocarbons such as natural gas and petroleum, and carbohydrates such as occur in plants and animals.

Chapter 3

Magnetic Observatories and Data Analysis

An observatory is a location used for observing terrestrial or celestial events. Astronomy, climatology/meteorology, geology, oceanography and volcanology are examples of disciplines for which observatories have been constructed. Historically, observatories were as simple as containing an astronomical sextant (for measuring the distance between stars) or Stonehenge (which has some alignments on astronomical phenomena).

Ground-Based Observatories

Ground-based observatories, located on the surface of Earth, are used to make observations in the radio and visible light portions of the electromagnetic spectrum. Most optical telescopes are housed within a dome or similar structure, to protect the delicate instruments from the elements. Telescope domes have a slit or other opening in the roof that can be opened during observing, and closed when the telescope is not in use. In most cases, the entire upper portion of the telescope dome can be rotated to allow the instrument to observe different sections of the night sky. Radio telescopes usually do not have domes.

For optical telescopes, most ground-based observatories are located far from major centers of population, to avoid the effects of light pollution. The ideal locations for modern observatories are sites that have dark skies, a large percentage of clear nights per year, dry air, and are at high elevations. At high elevations, the Earth's atmosphere is thinner thereby minimizing the effects of atmospheric turbulence and resulting in better astronomical "seeing". Sites that meet the above criteria for modern observatories include the southwestern United States, Hawaii, Canary Islands, the Andes, and high mountains in Mexico such as Sierra Negra. Major optical observatories include Mauna Kea Observatory and Kitt Peak National Observatory in the USA, Roque de los Muchachos Observatory in Spain, and Paranal Observatory in Chile.

Specific research study performed in 2009 shows that the best possible location for ground-based observatory on Earth is Ridge A – a place in the central part of Eastern Antarctica. This location provides the least atmospheric disturbances and best visibility.

Radio Observatories

Beginning in 1930s, radio telescopes have been built for use in the field of radio astronomy to observe the Universe in the radio portion of the electromagnetic spectrum. Such an instrument, or collection of instruments, with supporting facilities such as control centres, visitor housing, data reduction centers, and/or maintenance facilities are called *radio observatories*. Radio observatories are similarly located far from major population centers to avoid electromagnetic interference (EMI) from radio, TV, radar, and other EMI emitting devices, but unlike optical observatories, radio observatories can be placed in valleys for further EMI shielding. Some of the world's major radio observatories include the Socorro, in New Mexico, USA, Jodrell Bank in the UK, Arecibo, Puerto Rico, Parkes in New South Wales, Australia and Chajnantor in Chile.

Highest Astronomical Observatories

Since the mid-20th century, a number of astronomical observatories have been constructed at very high altitudes, above 4000–5000 m (13,000-16,000 ft). The largest and most notable of these is the Mauna Kea Observatory, located near the summit of a 4205 m (13,796 ft) volcano in Hawaii. The Chacaltaya Astrophysical Observatory in Bolivia, at 5230 m (17,160 ft), was the world's highest permanent astronomical observatory from the time of its construction during the 1940s until 2009. It has now been surpassed by the new University of Tokyo Atacama Observatory, an optical-infrared telescope on a remote 5640 m (18,500 ft) mountaintop in the Atacama Desert of Chile.

Space-Based Observatories

Space-based observatories are telescopes or other instruments that are located in outer space, many in orbit around the Earth. Space-based observatories can be used to observe astronomical objects at wavelengths of the electromagnetic spectrum that cannot penetrate the Earth's atmosphere and are thus impossible to observe using ground-based telescopes. The Earth's atmosphere is opaque to ultraviolet radiation, X-rays, and gamma rays and is partially opaque to infrared radiation so observations in these portions of the electromagnetic spectrum are best carried out from a location above the atmosphere of our planet.

Another advantage of space-based telescopes is that, because of their location above the Earth's atmosphere, their images are free from the effects of atmospheric turbulence that plague ground-based observations. As a result, the angular resolution of space telescopes such as the Hubble Space Telescope is often much smaller than a ground-based telescope with a similar aperture. However, all these advantages do come with a price. Space telescopes are much more expensive to build than ground-based telescopes. Due to their location, space telescopes are also extremely difficult to maintain. The Hubble Space Telescope was serviced by the Space Shuttle while many other space telescopes cannot be serviced at all.

Airborne Observatories

Airborne observatories have the advantage of height over ground installations, putting them above most of the Earth's atmosphere. But they also have an advantage over space telescopes – the instruments can be deployed, repaired, updated much more quickly and inexpensively. The Kuiper Airborne Observatory and the Stratospheric Observatory for Infrared Astronomy use airplanes to observe in the infrared, which is absorbed by water vapor in the atmosphere. Balloons for X-ray astronomy have been used in a variety of countries.

Colaba Observatory

Colaba Observatory was an astronomical, timekeeping, geomagnetic and meteorological observatory located on the Island of Colaba, Mumbai (Bombay), India.

The Colaba Observatory is located in Bombay and was built in 1826 by the East India Company for astronomical observations and time-keeping, with the purpose to provide support to British and other shipping which used the port of Bombay. The 165-year-old building served as office space for the Indian Institute of Geomagnetism. The recording of geomagnetism and meteorological observations was started at the observatory in 1841 by Arthur Bedford Orlebar, who was then Professor of Astronomy at Bombay's Elphinstone College. Magnetic measurements between the years of 1841 and 1845 were intermittent; following 1845 they became bi-hourly, then hourly. In 1845 Charles Brooke devised a self-recording photographic magnetometer with a light-source, a mirror for amplifying the magnet's movement, and a drum of photographic paper. This system gradually replaced the older manual method of taking eye-observations through vertical microscope which scanned the ends of a magnetic needle; the magnetic needle itself was suspended by a bunch of silk fibres.

The new photographic drum method ensured continuous recording of geomagnetic elements, and rapidly gained use all over the world. It came to Colaba in 1871, when Charles Chambers (later to become a Fellow of the Royal Society) held the Directorship. Colaba Observatory became more well known through his examination of geomagnetic measurements at Colaba, and his interpretation of the physics behind the phenomena. After his untimely death in Feb. 1896, the mantle of Directorship fell on the shoulders of Dr. Nanabhoy Ardeshir Framji Moos, the first Indian to hold this position.

With an Engineering degree from Poona, and a higher degree in Science from Edinburgh in Scotland, Dr. Nanabhoy Ardeshir Framji Moos saw to the efficient functioning of Colaba Observatory, regular analysis and interpretation of the measurements, and the starting of seismological observations. In 1900 Bombay decided to convert its fleet of horse-drawn trams to electric power for public transport. The electric trams would have vitiated the data from the Colaba magnetic observatory by generating electromagnetic noise.

Dr. Moos selected an alternate site at Alibag, located about 30 km directly to the south-east of Bombay. Alibag was located "far enough from Bombay to be free from the threatened electromagnetic noise, and yet near enough to retain the same geomagnetic characteristics". These aspects were checked out carefully over a 2-year period from 1904–1906, and then only was recording at Colaba discontinued, and the electric tram service started in Bombay. The entire building was made of hand-picked, non-magnetic, Porbandar sandstone, and magnetic recording is carried on in a room built with such good insulation that the variation in temperature within is just 10 °C over an entire day.

Of the entire Colaba-Alibag data, the French geomagnetician Pierre Noel Mayaud, had the following to say in 1973:

Finally, the (magnetic) records of Colaba and Alibag were found to form a beautiful series, beginning in 1871, and making up perhaps, the most complete collection of records in the world. Their quality and especially their regularity were particularly impressive, even in comparison with the Kew and Melbourne records.

Moos retired in 1919 after leading the Colaba-Alibag Observatories to worldwide renown. In 1910 he summarised the main findings from 50 years of geomagnetic measurement at the Colaba-Alibag Observatory over 1846–1905, in two volumes titled "Magnetic observations made at the Government Observatory, Bombay for the period 1846–1905. Parts I. and II." Of these volumes and of Colaba-Alibag's performance as a Geomagnetic Observatory, J.A. Fleming, a pioneer in Terrestrial Magnetism and Electricity, had the following to say in 1954:

The Golden Jubilee of the foundation of the Magnetic Observatory at Alibag (Mumbai), is a historic one in the field of Geomagnetism, and marks the long established application of India in an unparalleled series of magnetic recording of the phenomena, and publication of interpretative discussions of the accumulated data, as prepared under the direction of India's foremost investigator (N.A.F. Moos) in the two large volumes.

Despite over 1500 selected references in the field of geomagnetic research, Volume 3 of the Physics-of-the-Earth Series of the United States National Research Council, there is none which exhibits so wide and varied and intensive coverage of all the geomagnetic problems in the early 20th century.

Prof. K. R. Ramanathan, who assumed the Director position after Moos, and would later head the Physical Research Laboratory, Ahmedabad, said of his predecessor: "He was an ideal head of the observatory, always taking a deep interest in the welfare of his staff, and being held by them in great affection and esteem". Over the year 1919–1971, 17 Directors steered the Colaba-Alibag Observatories through avenues of meticulous and uninterrupted geomagnetic recordings, regular publishing of the data, and discussion of observations in scientific research journals.

In 1971 the Colaba-Alibag Observatories were converted into an autonomous research organisation called the Indian Institute of Geomagnetism. Until that point the Colaba-Alibag Observatories were part of the Indian Meteorological Department. Its headquarters continued to be in Mumbai, in the building constructed by John Curin in 1826, who was an astronomer for the East India Company. The first Director of the Indian Institute of Geomagnetism over 1971–1979 was

Prof. B. N. Bhargava. The proceeding Director was Prof. R. G. Rastogi over 1980–1989.

During the IGY-IGC years of 1957–1959, Prof. K. R. Ramanathan (a past Director of Colaba-Alibag), strongly advocated the setting up of magnetic observatories to examine the equatorial electrojet. The Trivandrum and Annamalainagar observatories were set up in November 1957, and were tended first under the Directorship of Mr. S. L. Malurkar, and then under Prof. P. R. Pisharoty.

Eighteen years elapsed before there was a need for further observatories along 75°E longitude meridian. The USSR sponsored "Project Geomagnetic Meridian" to serve their needs. Ujjain and Jaipur were consequently set up in July 1975, as wall as Shillong at 92°E longitude. In May 1977, Gulmarg, located very near the focus of the Sq. current system was started. In May 1991, the ninth observatory was started at Nagpur and then the observatories Vishakhapatnam, Pondicherry and Tirunelveli followed. Apart from these, a temporary station was run in the Andaman Islands in 1974 as support for the ONGC (Oil and Natural Gas Commission of India) in petroleum prospecting. Since 1979, an array of Gough-Reitzel magnetometers has operated at various sites in India for studies of the Earth's internal structure by examining electromagnetic induction within the earth. The Indian Institute of Geomagnetism currently operates Ten magnetic observatories.

Solid Earth Geomagnetism

Most of the Earth's geological and geophysical activity occurs because our planet is cooling to space, thereby inducing currents of cold sinking and hot rising material, otherwise known as convection. Convection in the Earth's mantle is the engine of plate tectonics and gives rise to the creation of ocean basins and continents; similarly, convection in the Earth's liquid-iron outer-core powers the geomagnetic field. Understanding how the Earth has been evolving in this way is one of the many aspects of studying the physics of the Earth's interior. Geophysical research is fundamentally multidisciplinary employ a variety of observational, experimental, and theoretical approaches to investigate the structure and dynamics of the Earth from atomic to global scale.

Seismology uses elastic waves generated by natural earthquakes as well as controlled sources to probe the internal structure of the Earth. Seismic wave propagation through the mantle, for example, is sensitive not only to the hot and cold regions that induce buoyant convection, but also to the fabric of mantle rock sheared by convection (known as seismic anisotropy). Research on seismic anisotropy, which yields insight into mantle flow, is complementary to laboratory work on rock deformation at deep-earth conditions.

The structure and dynamics of the crust and mantle is elucidated using temporary deployments of broadband seismometers in regions of tectonic interest, such as subduction zones. The seismic structure of crust imaged by active-source seismology, on the other hand, can constrain parental mantle processes such as convection and partial melting in the past and present, providing a robust geophysical perspective to igneous petrology.

Geophysics plays a major role in the understanding of the earth's interior, as well as in mineral and oil exploration. More often than not, the geophysical targets are not accessible for direct observation and hence the target is characterized by the manifestations of its physical properties, such as, the density (Gravity techniques) reflections at the boundaries (natural and artificial source seismic studies), temperature (Heat Flow studies), magnetic properties (Magnetic anomaly maps and palaeomagnetic techniques) and the electrical structure (Magnetotelluric and other electromagnetic techniques). These different studies are put together to identify the nature of the target. The major thrust areas of research under Solid Earth Geophysics are

Electromagnetic Induction Studies

Electromagnetic induction based tools such as magnetovariational and magnetotelluric techniques are used for imaging electrical conductivity structures in the interior of the Earth and studying their geophysical implications, and utility as geophysical exploration tools.

Tectonomagnetic Study

Monitoring of the total geomagnetic field in the vicinity of fault (Jabalpur area) to detect changes in the static part of geomagnetic field due to general tectonic activity and crustal stress accumulation with a view to understanding an earthquake mechanism.

Radio wave signals from the Global Positioning System (GPS) satellites and Synthetic Aperture Radar (SAR) Interferometry is used to study the deformation of Earth's crust in the Indian region.

The Standard Earth Model

Our direct knowledge of the earth's interior is minuscule. The earth has a radius of about 6370 km, but the deepest scientific borehole ever drilled is only 12 km. To put this in perspective: if the earth were reduced to a tabletop globe 50 centimetres (20 inches) in diameter, the portion accessible to direct observation through the deepest borehole would be the equivalent of a very thin skin less than 1 millimetre (0.04 inch) thick. In other words, scientists have barely scratched the surface of our globe.

Nevertheless, over the past 100 years or so, geoscientists have put together a detailed picture of the earth's interior based largely on *indirect* evidence — mainly the behaviour of seismic waves that travel through the earth. The earth's interior is believed to consist of several concentric spheres: an outer solid crust, averaging 7 km thick beneath oceans and 35 km beneath continents; a mainly solid mantle extending to a depth of 2900 km; an outer core of liquid iron extending to a depth of 5150 km; and an inner core of solid iron, with a radius of about 1220 km.

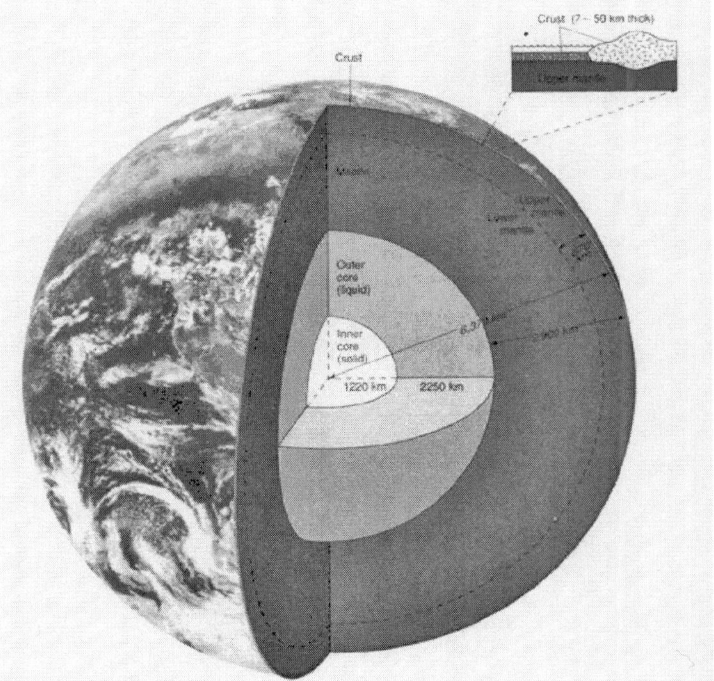

Figure: *The standard model of the earth's interior*

Whenever an earthquake occurs, seismic waves spread out from the focus in all directions. Three types of waves are distinguished: surface waves, body waves, and free oscillations (vibrations of the entire planet). Instead of travelling in straight lines, body waves are reflected and refracted (bent), depending on the density, pressure, and elasticity of the different layers of rock through which they pass. On the basis of the time taken by different types of waves from specific earthquakes to reach different parts of the earth's surface, seismologists try to work out the precise path the waves have taken, the changes in velocity they have undergone at different depths, and the density and composition of the earth at different depths. Nowadays this is done with the help of supercomputers.

Raypaths are immensely complex; waves may undergo multiple reflections and refractions, and their paths are further complicated by the fact that lateral heterogeneity exists at every depth in the earth. This is directly indicated by the scatter in seismic-wave arrival times at all distances from the source. Seismic tomography, which seeks to image the three-dimensional structure of the earth, provides indirect evidence of lateral variations of up to 10% in seismic velocity through the crust and mantle.

Scientists cannot even begin to interpret the hundreds of thousands of seismic records without making certain basic assumptions about the earth's interior. The main assumptions are that the earth's interior consists entirely of solid or liquid physical matter, and that temperature, pressure, and density increase with depth. These assumptions are generally believed to be self-evident.

At several depths in the earth, there appear to be discontinuities where the velocity of seismic waves changes abruptly. Such discontinuities are often transition zones rather than sharp boundaries, and vary in depth from place to place. The dominant boundary is that between the mantle and the core. Next in order of magnitude are the crust-mantle boundary (Mohorovicic discontinuity or Moho), the inner core-outer core boundary, and the mid-mantle discontinuities at depths of 400 and 670 km. The earth's core was 'discovered' in 1906, and its depth (about 2900 km) was determined by 1914. The Moho was 'discovered' in 1909, the inner core in 1936, and the 400- and 670-km discontinuities in the 1960s.

The thickness of the crust varies from 20 to 70 km beneath continents and from 5 to 15 km beneath oceans. As well as differing greatly in thickness, continental and oceanic crusts are said to have very different compositions: continental crust consists chiefly of granitic rock capped by sedimentary rocks, while ocean crust is believed to be composed largely of basalt and gabbro. At the crust-mantle boundary or Moho, seismic-wave velocities change abruptly, but there is no consensus on exactly why. No drillhole has yet penetrated to the Moho anywhere.

The Moho varies considerably in depth, sometimes several Mohos are stacked up, and in places there is no Moho at all. Sometimes it is flat, continuous, and oblivious to faults, while in other areas it is strongly influenced by overlying geological structures and jumps from one depth to another.

At the two main discontinuities in the mantle, rocks are widely believed to undergo pressure transformations to denser phases. The 670-km discontinuity marks the boundary between the upper and lower mantle; seismic waves increase suddenly in speed at this depth, and earthquakes essentially cease.

The mantle is thought to be composed of the dense, ultrabasic (ultramafic) rock peridotite. This is because lava sometimes contains fragments of peridotite and mountain-building processes sometimes bring up wedges of peridotite, and in both cases this rock is assumed to come from the mantle. V. Sánchez Cela disagrees and argues that many geological and geophysical phenomena can be better explained if the upper mantle is far more sialic (granitic) than currently believed.

The outer core is said to consist mainly of liquid iron, and the inner core of solid iron. The reasoning behind this is as follows. There are two main types of seismic body waves: P waves (compressional or longitudinal waves) and S waves (transverse or shear waves). P waves can travel through solids, liquids, and gases, while S waves can only travel through solids.

Seismic waves do not reach certain areas on the opposite side of the earth from a large earthquake. P waves spread out until, at 103° of arc (11,500 km) from the epicentre, they almost entirely disappear from seismograms. At more than 142° (15,500 km) from the epicentre, they reappear. The region in between is called the P-wave shadow zone. P waves are said to be missing in the shadow zone because they are refracted by the core.

The S-wave shadow zone is larger than the P-wave shadow zones; direct S waves are not recorded in the entire region more than 103° away from the epicentre. It therefore seems that S waves do not travel through the core at all, and this is interpreted to mean that it is liquid, or at least acts like a liquid.

The way P waves are refracted in the core is believed to indicate that there is a solid inner core. Although most of the earth's iron is supposed to be concentrated in the core, it is interesting to note that in the outer zones of the earth, iron levels *decrease with depth.*

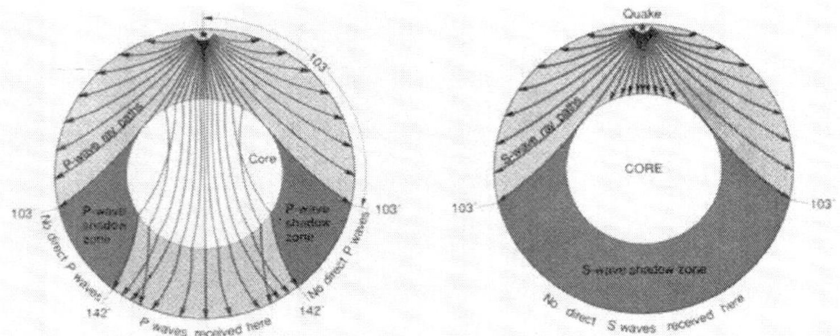

Figure: *P-wave and S-wave shadow zones.*

Seismologists sometimes draw contradictory conclusions from the same seismic data. For instance, two groups of geophysicists produced completely different pictures of the core-mantle boundary, where there are believed to be 'mountains' and 'valleys' as high or deep as 10 km. The two groups used virtually the same data but used different equations to process them [6]. Seismologists also disagree on the rate of rotation of the inner core: some say it is rotating faster than the rest of the planet, others that it is rotating more slowly, and yet others that it rotates at the same speed

It is becoming increasingly evident that the earth model presented by the reigning theory of plate tectonics is seriously flawed [8]. The rigid lithosphere, comprising the crust and uppermost mantle, is said to be fractured into several 'plates' of varying sizes, which move over a relatively plastic layer of partly molten rock known as the asthenosphere (or low-velocity zone). The lithosphere is said to average about 70 km thick beneath oceans and to be 100 to 250 km thick beneath continents. A powerful challenge to this model is posed by seismic tomography, which shows that the oldest parts of the continents have deep roots extending to depths of 400 to 600 km, and that the asthenosphere is essentially absent beneath them. Seismic research shows that even under the oceans there is no continuous asthenosphere, only disconnected asthenospheric lenses.

The more we learn about the crust and uppermost mantle, the more the models presented in geological textbooks are exposed as simplistic and unrealistic. The outermost layers of the earth have a highly complex, irregular, inhomogeneous structure; they are divided by faults into a mosaic of separate, jostling blocks of different shapes and sizes, generally a few hundred kilometres across, and of varying internal structure and strength. This fact, in conjunction with the existence of deep continental roots and the absence of a global asthenosphere, means that the notion of huge rigid plates moving thousands of kilometres across the earth is simply untenable. Continents are about as mobile as a brick in a wall!

The plate-tectonic hypothesis that the present oceans have formed by seafloor spreading since the early Mesozoic (within the last 200 million years) is also becoming increasingly implausible. Numerous far older continental rocks have been discovered in the oceans, along with 'anomalous' crustal types intermediate between standard 'continental' and 'oceanic' crust (e.g. plateaus, ridges, and rises), and the evidence for large (now submerged) continental landmasses in the present oceans continues to mount.

Deep Drilling Springs Surprises

How much faith can be put in the theories concerning the composition and density of rocks at different depths? The only place where the accuracy of scientific models can be tested directly is in the uppermost few kilometres of the crust. Although oil companies have drilled as deep as 8 km on land, they drill in sedimentary basins. The igneous and metamorphic basement, which averages 40 km thick and makes up most of the continental crust, has rarely been sampled deeper than 2 or 3 km.

The deepest borehole drilled for scientific purposes is located on the Kola Peninsula near Murmansk, Russia, in the northwestern part of the Baltic Shield. The drilling of the main borehole began in 1970, and a final depth of 12,262 metres was reached in 1994. The drilling of this and other deep and superdeep wells has produced one surprise after another, and the findings have been extremely embarrassing for earth scientists. One scientist commented: 'Every time we drill a hole we find the unexpected. That's exciting, but disturbing.' And a science reporter remarked: 'Kola revealed how far from truth scientific theory can roam.'

At the Kola hole, scientists expected to find 4.7 km of metamorphosed sedimentary and volcanic rock, then a granitic layer to a depth of 7 km (the 'Conrad discontinuity'), with a basaltic layer below it. The granite, however, appeared at 6.8 km and extends to more than 12 km; no basaltic layer was ever found! Seismic-reflection surveys, in which sound waves sent into the crust bounce back off contrasting rock types, have detected the Conrad discontinuity beneath all the continents, but the standard interpretation that it represents a change from granitic to basaltic rocks is clearly wrong. Metamorphic changes brought about by heat and pressure are now thought to be the most likely explanation.

Figure: *The 64-metre drill-rig enclosure over the 12-km-deep Kola borehole.*

The superdeep borehole at Oberpfälz, Germany, was expected to pass through a 3-to-5-km-thick nappe complex into a suture zone formed by a supposed continental collision. The borehole reached a final depth of 9101 m in 1994, but no evidence supporting the nappe concept was found. What the scientists did find was a series of nearly vertical folds that had failed to show up on seismic-reflection profiles.

Rock density is generally expected to increase with depth, as pressures rise. Results from the Kola hole indicated that densities did increase with depth initially, but at 4.5 km the drill encountered a sudden decrease in density, presumably due to increased porosity. The results also showed that increases in seismic velocity do not have to be caused by an increase in rock basicity.

The Soviet Minister of Geology reported that 'with increasing depth in the Kola hole, the expected increase in rock densities was therefore not recorded. Neither was any increase in the speed of seismic waves nor any other changes in the physical properties of the rocks detected. Thus the traditional idea that geological data obtained from the surface can be directly correlated with geological materials in the deep crust must be re-examined.'

The results of superdeep drilling show that seismic surveys of continental crust are being *systematically misinterpreted*. Much of the modelling of the earth's interior depends on the interpretation of seismic records. If these interpretations are wrong at depths of only a few kilometres, how much reliance can be placed on interpretations of the earth's structure at depths of hundreds or thousands of kilometres beneath the surface?!

Contrary to expectations, signs of rock alteration and mineralization were found as deep as 7 km in the Kola well. The hole intercepted a copper-nickel ore body almost 2 km below the level at which ore bodies were thought to disappear. In addition, hydrogen, helium, methane, and other gases, together with strongly mineralized waters were found circulating throughout the Kola hole. The presence of fractures open to fluid circulation at pressures of more than 3000 bars was entirely unexpected. The drillers at Oberpfälz discovered hot fluids in open fractures at 3.4 km. The brine was rich in potassium and twice as salty as ocean water, and its origin is a mystery.

Another surprise at the Kola hole was that lifeforms and fossils were discovered several kilometres down. Microscopic fossils were found at depths of 6.7 km. 24 species were identified among these microfossils, representing the envelopes or coverings of single-cell marine plants known as plankton. Unlike conventional shells of limestone or silica, these coverings were found to consist of carbon and nitrogen and had remained remarkably unaltered despite the high pressures and temperatures to which they had been subjected.

It is generally assumed that temperature increases with depth, reaching 1000°C at a depth of about 80 km, 4800°C at the core-mantle boundary, and 6900°C at the earth's centre. It is certainly true that mine shafts and oil drilling operations have indicated significant increases of temperature with depth. Indeed, superdeep drilling has shown that temperature increases with depth far more rapidly than predicted. In the Kola borehole, the temperature at 10-km depth was 180°C rather than the expected 100°C. Measurements revealed significant vertical variations in temperature gradient and heat-flux density along the borehole. Overall, the rate of temperature increase rose from 11° to 24°/km down to a depth of nearly 7 km, and then started to decline. Geologists recognise that the rate of temperature increase must drop off sharply at a certain depth as otherwise the mantle would be molten below about 100 km (even at the enormous pressures assumed to exist there), whereas seismic evidence indicates that it is solid.

The oceanic crust is commonly divided into three main layers: layer 1 consists of ocean-floor sediments and averages 0.5 km in thickness; layer 2 consists largely of basalt and is 1.0 to 2.5 km thick; and layer 3 is assumed to consist of gabbro and is about 5 km thick. A drillhole in the eastern Pacific Ocean has been reoccupied four times in a 12-year span, and has now reached a total depth of 2000 m below the seafloor. Seismic evidence suggested that the boundary between layers 2 and 3 would be found at a depth of about 1700 m, but the drill went well past that depth without finding the contact between the dikes of layer 2 and the expected gabbro of layer 3. Either the seismic interpretation or the model of layer 3's composition must be wrong [3].

As already mentioned, plate tectonics requires the crust beneath the oceans to be relatively young (no older than early Mesozoic), yet thousands of older rocks have been found in the world's oceans, and the geological and geophysical evidence already available strongly suggests that deeper ocean drilling will uncover more ancient sediments (including further remnants of continental landmasses) beneath the basaltic layer 2 that is currently — and conveniently — labelled 'basement'. This layer suggests that magma flooding was once ocean-wide, and studies of ocean sediments show that this activity was accompanied by progressive crustal subsidence in large sectors of the present oceans, beginning in the Jurassic.

Mass, Density, and Seismic Velocity

If the earth's interior were homogeneous, consisting of materials with the same properties throughout, seismic waves would travel in a straight line at a constant velocity. In reality, waves reach distant seismometers sooner than they would if the earth were homogeneous, and the greater the distance, the greater the acceleration. This implies that the waves arriving at the more distant stations have been travelling faster. Since seismic waves travel not only along the surface but also through the body of the earth, the earth's curvature will clearly result in stations more distant from an earthquake focus receiving waves that have passed through greater depths in the earth. From this it is inferred that the velocity of seismic waves increases with depth, due to changes in the properties of the earth's matter.

Seismic velocity in different media depends not just on the substance's density but also on its elastic properties (i.e. rigidity and incompressibility). In the case of solids and liquids, for instance, there is no correlation between sound-wave velocity and density [1]. Here are some examples involving metals:

Substance	*Density (g/cm^3)*	*Velocity of longitudinal waves (km/s)*
aluminium	2.7	6.42
zinc	7.1	4.21
iron	7.9	5.95
copper	8.9	4.76
nickel	8.9	6.04
gold	19.7	3.24

There *is* a correlation between density and seismic velocity in the case of gases: velocity *decreases* with increasing density due to the increased number of collisions.

According to the relevant equations,* the velocity of seismic waves will become *slower*, the *denser* the rocks through which they pass, if the rocks' elastic properties change in the same proportion as density. Since seismic waves accelerate with depth, this would imply that density *decreases*. However, scientists are convinced that the density of the rocks composing the earth's interior *increases* with depth. To get round this problem, they simply assume that the elastic properties change at a rate that more than compensates for the increase in density. As one textbook puts it:

Since the density of the Earth increases with depth you would expect the waves to slow down with increasing depth. Why, then, do both P- and S-waves speed up as they go deeper? This can only happen because the incompressibility and rigidity of the Earth increase faster with depth than density increases.

Thus geophysicists simply adjust the values for rigidity and incompressibility to fit in with their preconceptions regarding density and velocity distribution within the earth! In other words, their arguments are circular.

*P-wave velocity = square root of [(incompressibility + 4/3rigidity) divided by density]. S-wave velocity = square root of [rigidity divided by density]. In a fluid, rigidity vanishes and S waves cannot propagate at all.

Drilling results at the Kola borehole revealed significant heterogeneity in rock composition and density, seismic velocities, and other properties. Overall, rock porosity and pressure increased with depth, while density *decreased*, and seismic velocities showed no distinct trend. In the Oberpfälz pilot hole, too, density and seismic velocity showed no distinct trend with increasing depth. Many scientists believe that at greater depths, the presumed increase in pressures and temperatures will lead to greater homogeneity and that reality will approximate more closely to current models. But this is no more than a declaration of faith.

Scientists' conviction that density increases with depth is based on their belief that, due to the accumulating weight of the overlying rock, pressure must increase all the way to the earth's centre where it is believed to reach 3.5 million atmospheres (on the earth's surface the pressure is *one* atmosphere). They also believe that they know by how much rock density increases towards the earth's centre. This is because they think they have accurately determined the earth's mass (5.98 x 10^{24} kg) and therefore its *average* density (5.52 g/cm^3). Since the outermost crustal rocks — the only ones that can be sampled directly — have a density of only 2.75 g/cm^3, it follows that deeper layers of rock must be much denser. At the centre of the earth, density allegedly reaches 13.5 g/cm^3.

Pari Spolter Casts Doubt on this Model

About 71% of the earth's surface is covered by oceans at an average depth of 3795 m and mean density of 1.02 g cm^{-3}. The average thickness of the crust is 19 km and the mean crustal density is 2.75 g cm^{-3}. From studies of seismic wave travel time, geophysicists have outlined a layered structure in the interior of the earth. There is no accurate way currently known of estimating the density distribution from seismic data alone. To come up with a mean density of 5.5, earth models assuming progressively higher density values for the inner zones of the earth have been devised. . . .

Except for the ocean and the crust, direct measurements of the density of the inner layers of the earth are not available. This currently accepted Earth Model is inconsistent with the law of sedimentation in a centrifuge. The earth has been rotating for some 4.5 billion years. When it was first formed, the earth was in a molten state and was rotating faster than today. The highest density of matter should have migrated to the outer layers. Except for the inner core, . . . the density of the other layers of the earth should be less than 3 g cm^{-3}.

Also, heavy elements are rare in the universe. How could so much of materials with such low stellar abundances have concentrated in the earth's interior?

The figures given for the masses and densities of all planets, stars, etc. are *purely theoretical*; nobody has ever placed one on a balance and weighed it! The masses of celestial bodies can be calculated from what is known as Newton's form of Kepler's third law. Kepler's law states that the ratio of the cube of the mean distance (r) of each planet from the sun to the square of its period of revolution (T) is always the same number (r^3/T^2 = constant). Newton's version of this law assumes that r^3/T^2 is equal to the inert mass of the body multiplied by the gravitational constant divided by $4\pi^2$ ($GM = 4\pi^2 r^3/T^2$). Even if the conventional figures for the total mass and average density of the earth are correct, the prevailing earth model may still be very wrong since no one knows for certain what type of matter exists at the very centre of the earth.

The Devil's Dictionary defines gravitation as: 'The tendency of all bodies to approach one another with a strength proportioned to the quantity of matter they contain – the quantity of matter they contain being ascertained by the strength of their tendency to approach one another→ Such is the circular logic on which standard gravity theory is based. It should be borne in mind, however, that weight is always a relative measure, since one mass can only be weighed in relation to some other mass. The fact that observed artificial satellite speeds match predictions can be seen as evidence that the fundamentals of newtonian theory must be correct. On the other hand, the net gravitational force need not be directly proportional to inert mass, for there is plenty of evidence that characteristics such as spin and charge can modify a body's gravitational properties.

Deep Earthquakes

Most earthquakes are shallow, no deeper than 20-25 km, and occur when rocks snap and fracture under increasing stress. Earthquakes at much greater depths pose a major challenge to the standard earth model because below about 60 km, the rocks should be so hot and tightly compacted that they become ductile; instead of breaking catastrophically under stress, they should deform or flow plastically. Yet 30% of earthquakes occur at depths exceeding 70 km, and some have been recorded as far down as 700 km.

Most deep-focus earthquakes occur in Benioff zones; in plate-tectonic theory these deep-rooted fault zones are labelled 'subduction zones', where slabs of ocean lithosphere supposedly plunge into the earth's mantle (though there is abundant evidence contradicting this hypothesis). However, some deep earthquakes have shaken Romania and the Hindu Kush where there are no 'subduction zones'. A variety of mechanisms for deep earthquakes have been proposed, but they are all controversial.

The seismic radiation of deep earthquakes is similar to that of shallow earthquakes. It used to be said that deep-focus earthquakes were followed by fewer aftershocks than shallow ones, but there are indications that many of the aftershocks are simply difficult to detect, and that there is much more activity at such depths than is currently believed. The fact that deep earthquakes share many characteristics with shallow earthquakes suggests that they may be caused by similar mechanisms.

However, most earth scientists are incapable of entertaining the notion that the earth could be rigid at such depths. One exception is

E.A. Skobelin, who draws the logical conclusion that since deep-focus earthquakes cannot originate in plastic material but must be linked to some kind of stress in solid rock, the solid, rigid lithosphere must extend to depths of up to 700 km.

On 8 June 1994, one of the largest deep earthquakes of the 20th century, with a magnitude of 8.3 on the Richter scale, exploded 640 km beneath Bolivia. It caused the whole earth to ring like a bell for months on end; every 20 minutes or so, the entire planet expanded and contracted by a minute amount. A significant feature of the Bolivian earthquake was that it extended horizontally across a 30- by 50-km plane within the 'subducting slab'. This undermines the hypothesis that such quakes are caused by olivine within the 'cold' centre of a slab suddenly being transformed into spinel in a runaway reaction when the temperature rises above 600°C. It also undermines the theory that gravity increases with depth; if this were true, the motion of earthquakes at such depths should be nearly vertical. There appears to be something very wrong with scientific theories about what exists and what is happening deep within the earth.

The acceleration due to gravity is 9.8 m/s^2 at the earth's surface and the prevailing view is that it rises to a maximum of 10.4 m/s^2 at the core-mantle boundary (2900 km), before falling to zero at the earth's centre. But not all earth scientists agree. Skobelin argues that the normal, downwardly-directed gravitational force may be replaced by a reversed, upwardly-directed force at depths of 2700 to 4980 km, and that the widely-accepted figure of 3500 kilobars for the pressure at the earth's centre, may be an order of magnitude too high.

Earthquakes and volcanoes tend to concentrate along certain major fault lines in the earth's crust. The fact that heightened geological activity occurs along these 'plate boundaries' is sometimes hailed as one of the great successes of plate tectonics. However, it is precisely the high incidence of earthquake and volcanic activity that led geologists to label these belts as 'plate boundaries' in the first place! Plate tectonics sheds no light on earthquakes that happen within plates. Officer and Page state: 'We know very little about the mechanisms involved in such intraplate earthquakes, but [they sometimes] illustrate effects that one might expect from a gigantic internal explosion, odd as such a concept may appear'.

Thomas Gold has argued that, during its formation, the earth retained large quantities of hydrocarbons in its interior. He holds that various gases are sometimes released from depths of about 150 km, and when they invade the outer brittle layers of rock they weaken them by creating new fractures or reducing friction in existing faults, thereby causing or facilitating earthquakes. The emission of gases (e.g. methane) from the ground is already known to cause mud volcanoes on land, circular pockmarks on the ocean floor, and 'ice volcanoes' or pingos on ice fields. Hydrocarbons and hydrogen are also major components of the gases emitted during major volcanic eruptions.

Eyewitness accounts provide strong evidence that gas emissions also help to cause earthquakes in general, but nowadays scientists tend to ignore these 'subjective' accounts in favour of 'hard' seismic data. Eruptions, flames, roaring and hissing noises, sulphurous odours, hazes and fogs, asphyxiation, fountains of water and mud, vigorous bubbling in bodies of water — all these are observed today in conjunction with earthquakes, just as they were in past. On the basis of such evidence, the ancients held that the movement and eruption of subterranean 'air' (i.e. gases) caused volcanoes if they found an outlet, and otherwise generated earthquakes. Gold argues that this mechanism could explain deep earthquakes, since he believes that the mechanism of sudden rock shear cannot operate deep in the earth's interior. But as already noted, this belief may be wrong, and both mechanisms may apply at all depths.

Geomagnetism

Most earth scientists believe that, as well as having a high density, the earth's core, unlike the mantle, must be metallic in order to generate the geomagnetic field. According to the dynamo theory, fluid motion in the earth's outer core moves conducting material (liquid iron) across an already existing, weak magnetic field and generates an electric current. The electric current, in turn, produces a magnetic field that also interacts with the fluid motion to create a secondary magnetic field. Together, the two fields are stronger than the original and lie essentially along the earth's rotation axis.

The main characteristics of the geomagnetic field include short-term and long-term fluctuations in intensity, reversals of polarity at irregular intervals (ranging from tens of thousands to tens of millions of years), the 11° offset between the geomagnetic axis and spin axis, and the drift of the magnetic poles around the geographic poles in an estimated period of several thousand years. Scientists assume that the dynamo theory can explain these features, though a detailed understanding is lacking. There are competing dynamo models, and a great deal of fudging is required to get the numerical models to reproduce some of the features of the actual magnetic field.

To explain the offset between the earth's geomagnetic axis and the spin axis, some scientists maintain that the earth's overall field may be a combination of a central, dynamo-created dipole field, aligned with the rotation axis, and several variable dipole fields located in the outermost portions of the core. But other scientists argue that there is no physical mechanism to generate dipoles near the core's surface [2]. Some planets have even greater and more puzzling tilts between their magnetic and rotation axes: 46.8° in the case of Neptune, and 58.6° in the case of Uranus.

Even assuming that an outer core of liquid iron exists, there are major problems with the dynamo theory. Joseph Cater writes:

Scientists are somewhat vague as to how a magnetic field could extend 2,000 miles beyond an electric current. It requires a very powerful current to produce even relatively weak magnetic effects a very short distance above the flow. The electrical resistance of iron, at the alleged temperatures of the core, would be staggering. A steady flow of electricity requires constant potential differences. How are such potential differences produced and maintained in this hypothetical core?

The magnitude, width, and depth of such currents would have to be unbelievable to extend the magnetic field even a small fraction of the distance required, and the EMF [electromotive force] required to produce it would be even more incredible. Where could such an EMF come from? So far, scientists seem reluctant to explain this, especially since these currents are confined to a ball and would therefore follow closed paths.

V.N. Larin questions whether a mechanism exists to maintain strong electric currents in the earth's interior during its entire evolution, and argues that the very existence of active convection in the core is dubious. If convection is of thermal origin, then the source of heat in the iron core is incomprehensible. Another possibility is radioactivity, but no mechanism is known which might have segregated radioactive elements together with iron and nickel. Some scientists think that the heat source of convection may be the ongoing growth of the core. In this case, the heat would come from the potential energy of heavy particles settling in the gravity field, but this is unlikely to have lasted several billion years.

An alternative theory has been proposed by J.M. Herndon, who suggests that the earth's magnetic field is largely produced by electric currents generated by a self-sustaining nuclear fission reaction in a uranium (and thorium) subcore at the centre of the earth, having a density as high as 26 g/cm^3. However, the existence of such a subcore is entirely hypothetical.

Given their belief in the generation of magnetic fields by convection currents of electrically conducting liquid iron in a planet's core, scientists were puzzled by the discovery that the Moon and Mercury had significant magnetic fields, since the Moon's core is believed to be entirely solid and Mercury's core nearly so. Venus is believed to have an entirely liquid core and was expected to possess a strong magnetic field, but no significant self-generated field has been detected. The magnetic fields of Jupiter and Saturn are believed to be generated by electric currents within a layer of liquid metallic hydrogen inside them, while the fields of Neptune and Uranus are thought to be produced in their superheated liquid mantles — but all this is little better than guesswork. Clearly the present dynamo theory cannot explain the magnetic fields detected around some asteroids.

Atmosphere of Earth

The atmosphere of Earth is a layer of gases surrounding the planet Earth that is retained by Earth's gravity. The atmosphere protects life on Earth by absorbing ultraviolet solar radiation, warming the surface through heat retention (greenhouse effect), and reducing temperature extremes between day and night (the diurnal temperature variation).

The common name given to the atmospheric gases used in breathing and photosynthesis is air. By volume, dry air contains 78.09% nitrogen, 20.95% oxygen, 0.93% argon, 0.039% carbon dioxide, and small amounts of other gases. Air also contains a variable amount of water vapor, on average around 1%. Although air content and atmospheric pressure vary at different layers, air suitable for the survival of terrestrial plants and terrestrial animals currently is only known to be found in Earth's troposphere and artificial atmospheres.

The atmosphere has a mass of about 5.15×10^{18} kg, three quarters of which is within about 11 km (6.8 mi; 36,000 ft) of the surface. The atmosphere becomes thinner and thinner with increasing altitude, with no definite boundary between the atmosphere and outer space. The Kármán line, at 100 km (62 mi), or 1.57% of Earth's radius, is often used as the border between the atmosphere and outer space. Atmospheric effects become noticeable during atmospheric re-entry of spacecraft at an altitude of around 120 km (75 mi). Several layers can be distinguished in the atmosphere, based on characteristics such as temperature and composition.

Composition

Air is mainly composed of nitrogen, oxygen, and argon, which together constitute the major gases of the atmosphere. Water vapor accounts for roughly 0.25% of the atmosphere by mass. The concentration of water vapor (a greenhouse gas) varies significantly from around 10 ppmv in the coldest portions of the atmosphere to as much as 5% by volume in hot, humid air masses, and concentrations of other atmospheric gases are typically provided for dry air without any water vapor.

The remaining gases are often referred to as trace gases, among which are the greenhouse gases such as carbon dioxide, methane, nitrous oxide, and ozone. Filtered air includes trace amounts of many other chemical compounds.

Many substances of natural origin may be present in locally and seasonally variable small amounts as aerosols in an unfiltered air sample, including dust of mineral and organic composition, pollen and spores, sea spray, and volcanic ash. Various industrial pollutants also may be present as gases or aerosols, such as chlorine (elemental or in compounds), fluorine compounds and elemental mercury vapor. Sulfur compounds such as hydrogen sulfide and sulfur dioxide (SO_2) may be derived from natural sources or from industrial air pollution.

Major constituents of dry air, by volume			
Gas		***Volume***[(A)]	
Name	***Formula***	***in ppmv***[(B)]	***in %***
Nitrogen	N_2	780,840	78.084
Oxygen	O_2	209,460	20.946
Argon	Ar	9,340	0.9340
Carbon dioxide	CO_2	397	0.0397
Neon	Ne	18.18	0.001818
Helium	He	5.24	0.000524
Methane	CH_4	1.79	0.000179
Not included in above dry atmosphere:			
Water vapor[(C)]	H_2O	10–50,000[(D)]	0.001%–5%[(D)]

notes:

(A) volume fraction is equal to mole fraction for ideal gas only, also see volume (thermodynamics)

(B) ppmv: parts per million by volume

(C) Water vapor is about 0.25% *by mass* over full atmosphere

(D) Water vapor strongly varies locally

Structure of the Atmosphere

Principal Layers: In general, air pressure and density decrease with altitude in the atmosphere. However, temperature has a more complicated profile with altitude, and may remain relatively constant or even increase with altitude in some regions. Because the general pattern of the temperature/altitude profile is constant and recognisable through means such as balloon soundings, the temperature behaviour provides a useful metric to distinguish between atmospheric layers. In this way, Earth's atmosphere can be divided (called atmospheric stratification) into five main layers. From highest to lowest, these layers are:

- Exosphere: >700 km (>440 miles)
- Thermosphere: 80 to 700 km (50 to 440 miles)
- Mesosphere: 50 to 80 km (31 to 50 miles)
- Stratosphere: 12 to 50 km (7 to 31 miles)
- Troposphere: 0 to 12 km (0 to 7 miles)

Earth's Atmosphere: Lower 4 layers of the atmosphere in 3 dimensions as seen diagonally from above the exobase. Layers drawn to scale, objects within the layers are not to scale. Aurorae shown here at the bottom of the thermosphere can actually form at any altitude in this atmospheric layer

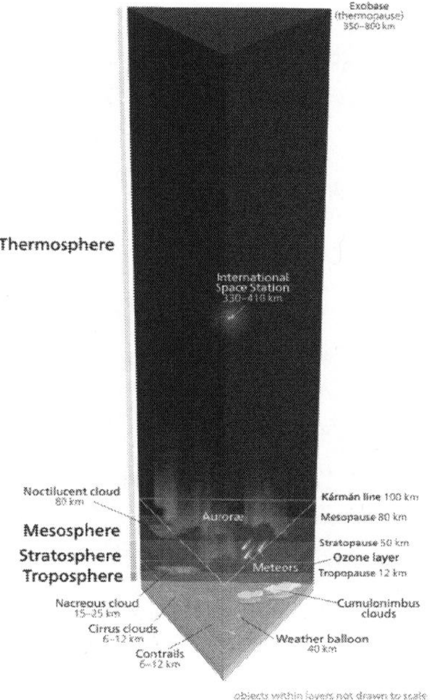

Exosphere

The exosphere is the outermost layer of Earth's atmosphere (i.e. the upper limit of the atmosphere). It extends from the exobase, which is located at the top of the thermosphere at an altitude of about 700 km above sea level, to about 10,000 km (6,200 mi; 33,000,000 ft). The exosphere merges with the emptiness of outer space, where there is no atmosphere.

This layer is mainly composed of extremely low densities of hydrogen, helium and several heavier molecules including nitrogen, oxygen and carbon dioxide closer to the exobase. The atoms and molecules are so far apart that they can travel hundreds of kilometers without colliding with one another. Thus, the exosphere no longer behaves like a gas, and the particles constantly escape into space. These free-moving particles follow ballistic trajectories and may migrate in and out of the magnetosphere or the solar wind.

The exosphere is located too far above Earth for any meteorological phenomena to be possible. However, the aurora borealis and aurora australis sometimes occur in the lower part of the exosphere, where they overlap into the thermosphere. The exosphere contains most of the satellites orbiting Earth.

Thermosphere

The thermosphere is the second-highest layer of Earth's atmosphere. It extends from the mesopause (which separates it from the mesosphere) at an altitude of about 80 km (50 mi; 260,000 ft) up to the thermopause at an altitude range of 500–1000 km (310–620 mi; 1,600,000–3,300,000 ft). The height of the thermopause varies considerably due to changes in solar activity. Since the thermopause lies at the lower boundary of the exosphere, it is also referred to as the exobase. The lower part of the thermosphere, from 80 to 550 kilometres (50 to 342 mi) above Earth's surface, contains the ionosphere.

This atmospheric layer undergoes a gradual increase in temperature with height. Unlike the stratosphere, wherein a temperature inversion is due to the absorption of radiation by ozone, the inversion in the thermosphere occurs due to the extremely low density of its molecules. The temperature of this layer can rise as high as 1500 °C (2700 °F), though the gas molecules are so far apart that its temperature in the usual sense is not very meaningful. The air is so rarefied that an individual molecule (of oxygen, for example) travels an average of 1 kilometre (0.62 mi; 3300 ft) between collisions with other molecules. Even though the thermosphere has a very high proportion of molecules with immense amounts of energy, the thermosphere would still feel extremely cold to a human in direct contact because the total energy of its relatively few number of molecules is incapable of transferring an adequate amount of energy to the skin of a human. In other words, a person would not feel warm because of the thermosphere's extremely low pressure.

This layer is completely cloudless and free of water vapor. However non-hydrometeorological phenomena such as the aurora borealis and aurora australis are occasionally seen in the thermosphere. The International Space Station orbits in this layer, between 320 and 380 km (200 and 240 mi).

Mesosphere

The mesosphere is the third highest layer of Earth's atmosphere, occupying the region above the stratosphere and below the thermosphere. It extends from the stratopause at an altitude of about 50 km (31 mi; 160,000 ft) to the mesopause at 80–85 km (50–53 mi; 260,000–280,000 ft) above sea level.

Temperatures drop with increasing altitude to the mesopause that marks the top of this middle layer of the atmosphere. It is the coldest place on Earth and has an average temperature around –85 °C (–120 °F; 190 K).

Just below the mesopause, the air is so cold that even the very scarce water vapor at this altitude can be sublimated into polar-mesospheric noctilucent clouds. These are highest clouds in the atmosphere and may be visible to the naked eye if sunlight reflects off them about an hour or two after sunset or a similar length of time before sunrise. They are most readily visible when the Sun is around 4 to 16 degrees below the horizon.

A type of lightning referred to as either sprites or ELVES, occasionally form far above tropospheric thunderclouds. The mesosphere is also the layer where most meteors burn up upon atmospheric entrance. It is too high above Earth to be accessible to jet-powered aircraft, and too low to support satellites and orbital or sub-orbital spacecraft. The mesosphere is mainly accessed by rocket-powered aircraft and unmanned sounding rockets.

Stratosphere

The stratosphere is the second-lowest layer of Earth's atmosphere. It lies above the troposphere and is separated from it by the tropopause. This layer extends from the top of the troposphere at roughly 12 km

(7.5 mi; 39,000 ft) above Earth's surface to the stratopause at an altitude of about 50 to 55 km (31 to 34 mi; 164,000 to 180,000 ft).

The atmospheric pressure at the top of the stratosphere is roughly 1/1000 the pressure at sea level. It contains the ozone layer, which is the part of Earth's atmosphere that contains relatively high concentrations of that gas. The stratosphere defines a layer in which temperatures rise with increasing altitude. This rise in temperature is caused by the absorption of ultraviolet radiation (UV) radiation from the Sun by the ozone layer, which restricts turbulence and mixing. Although the temperature may be –60 °C (–76 °F; 210 K) at the tropopause, the top of the stratosphere is much warmer, and may be near 0 °C.

The stratospheric temperature profile creates very stable atmospheric conditions, so the stratosphere lacks the weather-producing air turbulence that is so prevalent in the troposphere. Consequently, the stratosphere is almost completely free of clouds and other forms of weather. However, polar stratospheric or nacreous clouds are occasionally seen in the lower part of this layer of the atmosphere where the air is coldest. This is the highest layer that can be accessed by jet-powered aircraft.

Troposphere

The troposphere is the lowest layer of Earth's atmosphere. It extends from Earth's surface to an average height of about 12 km, although this altitude actually varies from about 9 km (30,000 ft) at the poles to 17 km (56,000 ft) at the equator, with some variation due to weather. The troposphere is bounded above by the tropopause, a boundary marked by stable temperatures.

Although variations do occur, the temperature usually declines with increasing altitude in the troposphere because the troposphere is mostly heated through energy transfer from the surface. Thus, the lowest part of the troposphere (i.e. Earth's surface) is typically the warmest section of the troposphere. This promotes vertical mixing (hence the origin of its name in the Greek word τρüπïò, *tropos*, meaning "turn"). The troposphere contains roughly 80% of the mass of Earth's atmosphere. The troposphere is denser than all its overlying atmospheric layers because a larger atmospheric weight sits on top of the troposphere and causes it to be most severely compressed. Fifty percent of the total mass of the atmosphere is located in the lower 5.6 km (18,000 ft) of the troposphere. It is primarily composed of nitrogen (78%) and oxygen (21%) with only small concentrations of other trace gases.

Nearly all atmospheric water vapor or moisture is found in the troposphere, so it is the layer where most of Earth's weather takes place. It has basically all the weather-associated cloud genus types generated by active wind circulation, although very tall cumulonimbus thunder clouds can penetrate the tropopause from below and rise into the lower part of the stratosphere.

Most conventional aviation activity takes place in the troposphere, and it is the only layer that can be accessed by propeller-driven aircraft.

Other Layers

Within the five principal layers which are largely determined by temperature, several secondary layers may be distinguished by other properties:

- The ozone layer is contained within the stratosphere. In this layer ozone concentrations are about 2 to 8 parts per million, which is much higher than in the lower atmosphere but still very small compared to the main components of the atmosphere. It is mainly located in the lower portion of the stratosphere from about 15–35 km (9.3–21.7 mi; 49,000–115,000 ft), though the thickness varies seasonally and geographically. About 90% of the ozone in our atmosphere is contained in the stratosphere.
- The ionosphere is a region of the atmosphere that is ionized by solar radiation. It is responsible for auroras. During daytime hours, it stretches from 50 to 1,000 km (31 to 621 mi; 160,000 to 3,280,000 ft) and includes the mesosphere, thermosphere, and parts of the exosphere.

  However, ionization in the mesosphere largely ceases during the night, so auroras are normally seen only in the thermosphere and lower exosphere. The ionosphere forms the inner edge of the magnetosphere. It has practical importance because it influences, for example, radio propagation on Earth.
- The homosphere and heterosphere are defined by whether the atmospheric gases are well mixed. The surfaced-based homosphere includes the troposphere, stratosphere, mesosphere, and the lowest part of the thermosphere, where the chemical composition of the atmosphere does not depend on molecular weight because the gases are mixed by turbulence. This relatively homogeneous layer ends at the *turbopause* which is found at about 100 km (62 mi; 330,000 ft), which places it about 20 km (12 mi; 66,000 ft) above the mesopause.

Above this altitude lies the heterosphere which includes the exosphere and most of the thermosphere. Here the chemical composition varies with altitude. This is because the distance that particles can move without colliding with one another is large compared with the size of motions that cause mixing. This allows the gases to stratify by molecular weight, with the heavier ones such as oxygen and nitrogen present only near the bottom of the heterosphere. The upper part of the heterosphere is composed almost completely of hydrogen, the lightest element.

- The planetary boundary layer is the part of the troposphere that is closest to Earth's surface and is directly affected by it, mainly through turbulent diffusion. During the day the planetary boundary layer usually is well-mixed, whereas at night it becomes stably stratified with weak or intermittent mixing. The depth of the planetary boundary layer ranges from as little as about 100 meters on clear, calm nights to 3000 m or more during the afternoon in dry regions.

The average temperature of the atmosphere at the surface of Earth is 14 °C (57 °F; 287 K) or 15 °C (59 °F; 288 K), depending on the reference.

Physical Properties

Pressure and Thickness

The average atmospheric pressure at sea level is defined by the International Standard Atmosphere as 101325 pascals (760.00 Torr; 14.6959 psi; 760.00 mmHg). This is sometimes referred to as a unit of standard atmospheres (atm). Total atmospheric mass is 5.1480×10^{18} kg (1.135×10^{19} lb), about 2.5% less than would be inferred from the average sea level pressure and Earth's area of 51007.2 megahectares, this portion being displaced by Earth's mountainous terrain. Atmospheric pressure is the total weight of the air above unit area at the point where the pressure is measured. Thus air pressure varies with location and weather.

If the entire mass of the atmosphere had a uniform density from sea level, it would terminate abruptly at an altitude of 8.50 km (27,900 ft). It actually decreases exponentially with altitude, dropping by half every 5.6 km (18,000 ft) or by a factor of 1/e every 7.64 km (25,100 ft), the average scale height of the atmosphere below 70 km (43 mi; 230,000 ft). However, the atmosphere is more accurately modelled with a customized equation for each layer that takes gradients of temperature, molecular composition, solar radiation and gravity into account.

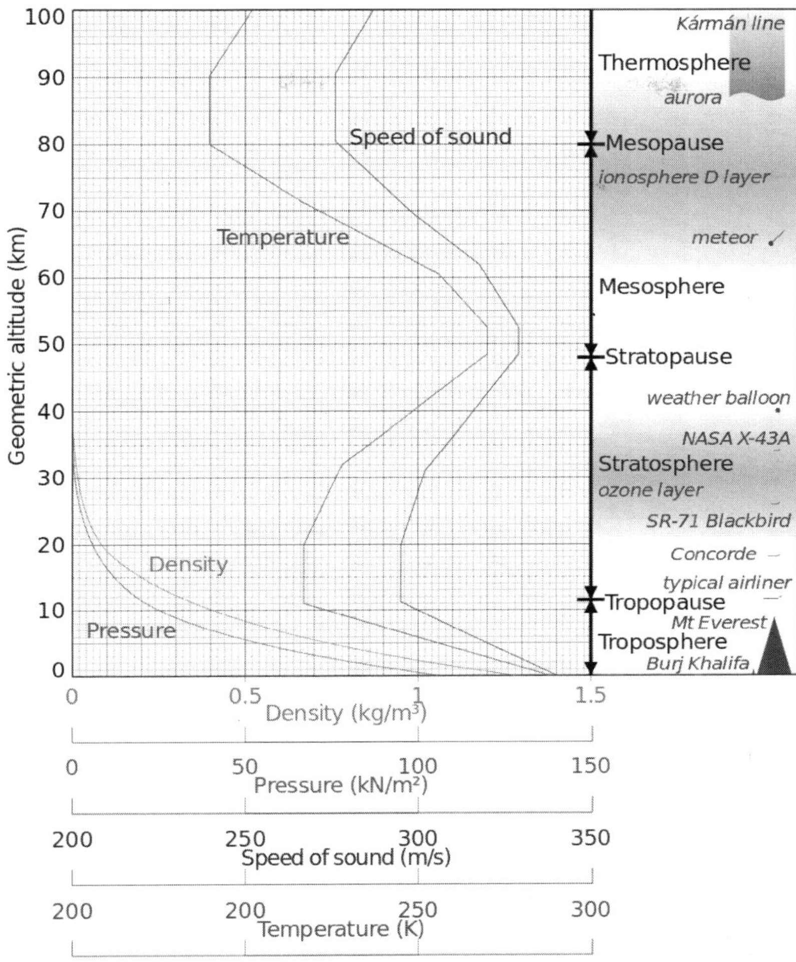

Figure: *Comparison of the 1962 US Standard Atmosphere graph of geometric altitude against air density, pressure, the speed of sound and temperature with approximate altitudes of various objects.*

In summary, the mass of Earth's atmosphere is distributed approximately as follows:

- 50% is below 5.6 km (18,000 ft).
- 90% is below 16 km (52,000 ft).
- 99.99997% is below 100 km (62 mi; 330,000 ft), the Kármán line. By international convention, this marks the beginning of space where human travelers are considered astronauts.

By comparison, the summit of Mt. Everest is at 8,848 m (29,029 ft); commercial airliners typically cruise between 10 km (33,000 ft) and

13 km (43,000 ft) where the thinner air improves fuel economy; weather balloons reach 30.4 km (100,000 ft) and above; and the highest X-15 flight in 1963 reached 108.0 km (354,300 ft).

Even above the Kármán line, significant atmospheric effects such as auroras still occur. Meteors begin to glow in this region though the larger ones may not burn up until they penetrate more deeply. The various layers of Earth's ionosphere, important to HF radio propagation, begin below 100 km and extend beyond 500 km. By comparison, the International Space Station and Space Shuttle typically orbit at 350–400 km, within the F-layer of the ionosphere where they encounter enough atmospheric drag to require reboosts every few months. Depending on solar activity, satellites can experience noticeable atmospheric drag at altitudes as high as 700–800 km.

Temperature and Speed of Sound

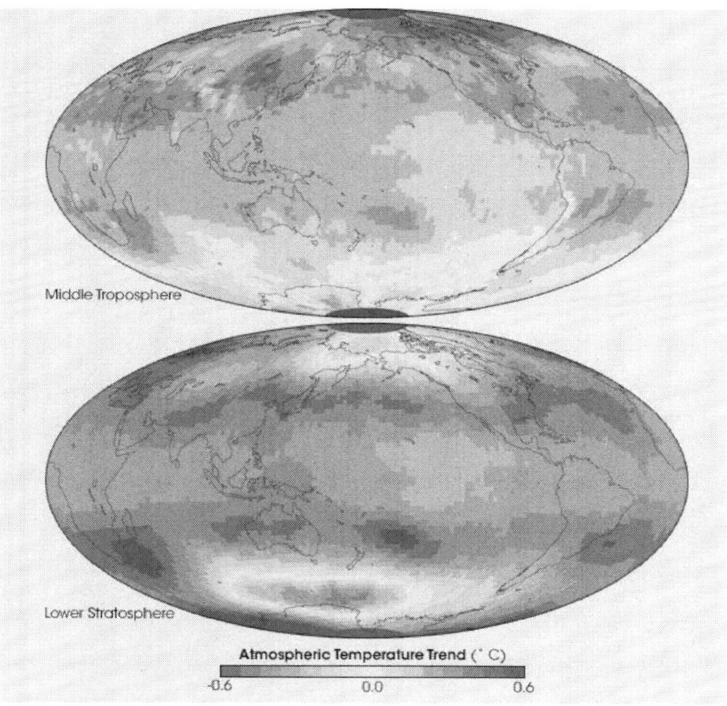

Figure: *These images show temperature trends in two thick layers of the atmosphere as measured by a series of satellite-based instruments between January 1979 and December 2005. The measurements were taken by Microwave Sounding Units and Advanced Microwave Sounding Units flying on a series of National Oceanic and Atmospheric Administration (NOAA) weather satellites. The instruments record microwaves emitted from oxygen molecules in the atmosphere. Source:*

The division of the atmosphere into layers mostly by reference to temperature is discussed above. Temperature decreases with altitude starting at sea level, but variations in this trend begin above 11 km, where the temperature stabilizes through a large vertical distance through the rest of the troposphere. In the stratosphere, starting above about 20 km, the temperature increases with height, due to heating within the ozone layer caused by capture of significant ultraviolet radiation from the Sun by the dioxygen and ozone gas in this region. Still another region of increasing temperature with altitude occurs at very high altitudes, in the aptly-named thermosphere above 90 km.

Because in an ideal gas of constant composition the speed of sound depends only on temperature and not on the gas pressure or density, the speed of sound in the atmosphere with altitude takes on the form of the complicated temperature profile, and does not mirror altitudinal changes in density or pressure.

Density and Mass

The density of air, ρ, is the mass per unit volume of Earth's atmosphere. Air density, like air pressure, decreases with increasing altitude. It also changes with variation in temperature or humidity. At sea level and at 15 °C, air has a density of approximately 1.225 kg/m³ (0.001225 g/cm³, 0.0023769 slug/ft³, 0.0765 lbm/ft³) according to ISA (International Standard Atmosphere).

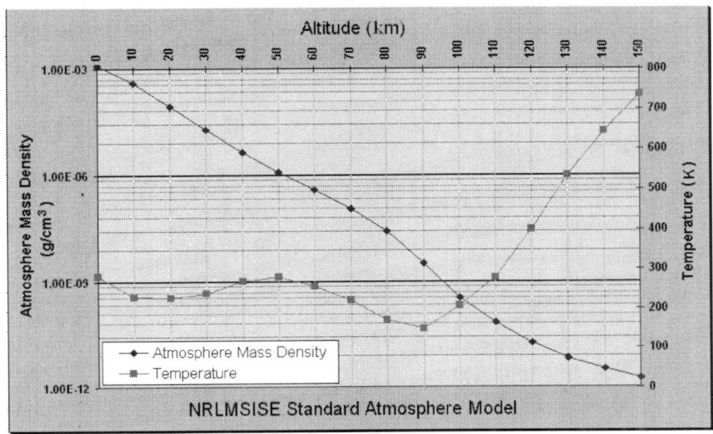

Figure: *Temperature and mass density against altitude from the NRLMSISE-00 standard atmosphere model (the eight dotted lines in each "decade" are at the eight cubes 8, 27, 64, ..., 729)*

The air density is a property used in many branches of science as aeronautics; gravimetric analysis; the air-conditioning industry; atmospheric research and meteorology; the agricultural engineering in their modelling and tracking of Soil-Vegetation-Atmosphere-Transfer (SVAT) models; and the engineering community that deals with compressed air from industry utility, heating, dry and cooling process in industry like a cooling towers, vacuum and deep vacuum processes, high pressure processes, the gas and light oil combustion processes that power our turbines airplanes, gas turbine-powered generators and heating furnaces, and air conditioning from deep mines to space capsules.

Depending on the measuring instruments, use, area of expertise and necessary rigor of the result different calculation criteria and sets of equations for the calculation of the density of air are used. This topic are some examples of calculations with the main variables involved, the amounts presented throughout these examples are properly referenced usual values, different values can be found in other references depending on the criteria used for the calculation. Furthermore we must pay attention to the fact that air is a mixture of gases and the calculation always simplify, to a greater or lesser extent, the properties of the mixture and the values for the composition according to the criteria of calculation.

The density of air at sea level is about 1.2 kg/m^3 (1.2 g/L). Density is not measured directly but is calculated from measurements of temperature, pressure and humidity using the equation of state for air (a form of the ideal gas law). Atmospheric density decreases as the altitude increases. This variation can be approximately modelled using the barometric formula. More sophisticated models are used to predict orbital decay of satellites.

The average mass of the atmosphere is about 5 quadrillion (5×10^{15}) tonnes or 1/1,200,000 the mass of Earth. According to the American National Center for Atmospheric Research, "The total mean mass of the atmosphere is 5.1480×10^{18} kg with an annual range due to water vapor of 1.2 or 1.5×10^{15} kg depending on whether surface pressure or water vapor data are used; somewhat smaller than the previous estimate. The mean mass of water vapor is estimated as 1.27×10^{16} kg and the dry air mass as $5.1352 \pm 0.0003\times10^{18}$ kg."

Density of Air Variables

Temperature and Pressure

The density of dry air can be calculated using the ideal gas law, expressed as a function of temperature and pressure:

$$\rho = \frac{p}{R_{\text{specific}} T}$$

where:

ρ = air density

p = absolute pressure

T = absolute temperature

$R_{specific}$ = specific gas constant for dry air

The specific gas constant for dry air is 287.058 J/(kg·K) in SI units, and 53.35 (ft·lb_f)/(lb_m·°R) in United States customary and Imperial units. This quantity may vary slightly depending on the molecular composition of air at a particular location.

Therefore:

- At IUPAC standard temperature and pressure (0 °C and 100 kPa), dry air has a density of 1.2754 kg/m^3.
- At 20 °C and 101.325 kPa, dry air has a density of 1.2041 kg/m^3.
- At 70 °F and 14.696 psi, dry air has a density of 0.074887lb_m/ft^3.

The following table illustrates the air density–temperature relationship at 1 atm or 101.325 kPa:

Effect of temperature on properties of air			
Temperature T (°C)	*Speed of sound c (m·s^{-1})*	*Density of air ρ (kg·m^{-3})*	*Characteristic specific acoustic impedance z_0 (Pa·m^{-1}·s)*
+35	351.88	1.1455	403.2
+30	349.02	1.1644	406.5
+25	346.13	1.1839	409.4
+20	343.21	1.2041	413.3
+15	340.27	1.2250	416.9
+10	337.31	1.2466	420.5
+5	334.32	1.2690	424.3
0	331.30	1.2922	428.0
–5	328.25	1.3163	432.1
–10	325.18	1.3413	436.1
–15	322.07	1.3673	440.3
–20	318.94	1.3943	444.6
–25	315.77	1.4224	449.1

Humidity (Water Vapor)

The addition of water vapor to air (making the air humid) reduces the density of the air, which may at first appear counter-intuitive. This occurs because the molar mass of water (18 g/mol) is less than the molar mass of dry air (around 29 g/mol).

For any gas, at a given temperature and pressure, the number of molecules present is constant for a particular volume. So when water molecules (water vapor) are added to a given volume of air, the dry air molecules must decrease by the same number, to keep the pressure or temperature from increasing. Hence the mass per unit volume of the gas (its density) decreases.

The density of humid air may be calculated as a mixture of ideal gases. In this case, the partial pressure of water vapor is known as the vapor pressure. Using this method, error in the density calculation is less than 0.2% in the range of –10 °C to 50 °C. The density of humid air is found by:

$$\rho_{\text{humid air}} = \frac{p_d}{R_d T} + \frac{p_v}{R_v T} = \frac{p_d M_d + p_v M_v}{RT}$$

where:

$\rho_{\text{humid air}}$ = Density of the humid air (kg/m^3)

p_d = Partial pressure of dry air (Pa)

R_d = Specific gas constant for dry air, 287.058 J/(kg·K)

T = Temperature (K)

p_v = Pressure of water vapor (Pa)

R_v = Specific gas constant for water vapor, 461.495 J/(kg·K)

M_d = Molar mass of dry air, 0.028964 kg/mol

M_v = Molar mass of water vapor, 0.018016 kg/mol

R = Universal gas constant, 8.314 J/(K·mol)

The vapor pressure of water may be calculated from the saturation vapor pressure and relative humidity. It is found by:

$$p_v = \phi p_{\text{sat}}$$

where:

p_v = Vapor pressure of water

ϕ = Relative humidity

p_{sat} = Saturation vapor pressure

The saturation vapor pressure of water at any given temperature is the vapor pressure when relative humidity is 100%. One formula used to find the saturation vapor pressure is:

$$p_{\text{sat}} = 6.1078 \times 10^{\frac{7.5T}{T+237.3}}$$

where T = is in degrees C.

Note:

- This equation will give the result of pressure in hPa (100 Pa, equivalent to the older unit millibar, 1 mbar = 0.001 bar = 0.1 kPa)

The partial pressure of dry air p_d is found considering partial pressure, resulting in:

$$p_d = p - p_v$$

Where p simply denotes the observed absolute pressure.

Altitude

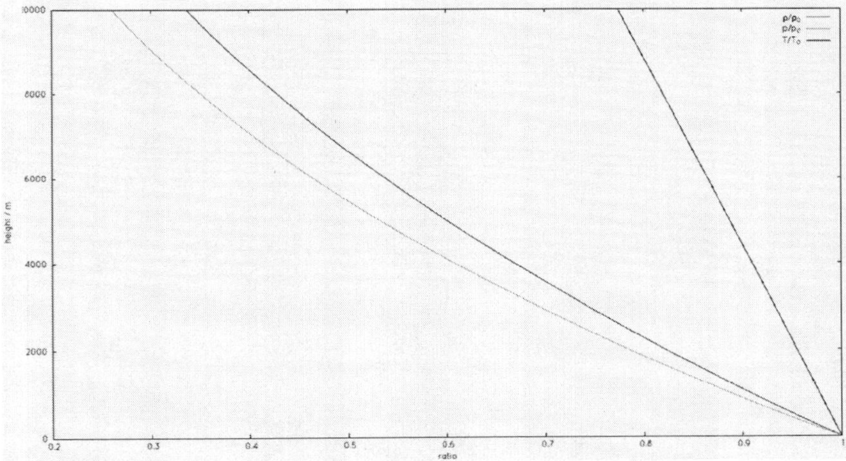

Figure: *Standard Atmosphere:* p_0*=101.325 kPa,* T_0*=288.15 K,* ρ *=1.225 [[kg/m³]]*

To calculate the density of air as a function of altitude, one requires additional parameters.

They are listed below, along with their values according to the International Standard Atmosphere, using for calculation the universal gas constant instead of the air specific constant:

p_0 = sea level standard atmospheric pressure, 101.325 kPa

T_0 = sea level standard temperature, 288.15 K

g = earth-surface gravitational acceleration, 9.80665 m/s^2

L = temperature lapse rate, 0.0065 K/m

R = ideal (universal) gas constant, 8.31447 J/(mol·K)

M = molar mass of dry air, 0.0289644 kg/mol

Temperature at altitude h meters above sea level is approximated by the following formula (only valid inside the troposphere):

$$T = T_0 - Lh$$

The pressure at altitude h is given by:

$$p = p_0 \left(1 - \frac{Lh}{T_0}\right)^{\frac{gM}{RL}}$$

Density can then be calculated according to a molar form of the ideal gas law:

$$\rho = \frac{pM}{RT}$$

where:

M = molar mass

R = ideal gas constant

T = absolute temperature

p = absolute pressure must be in Pa and not the kPa above.

Optical Properties

Solar radiation (or sunlight) is the energy Earth receives from the Sun. Earth also emits radiation back into space, but at longer wavelengths that we cannot see. Part of the incoming and emitted radiation is absorbed or reflected by the atmosphere.

Scattering

When light passes through our atmosphere, photons interact with it through *scattering*. If the light does not interact with the atmosphere, it is called *direct radiation* and is what you see if you were to look directly at the Sun. *Indirect radiation* is light that has been scattered in the atmosphere. For example, on an overcast day when you cannot see your shadow there is no direct radiation reaching you, it has all been scattered.

As another example, due to a phenomenon called Rayleigh scattering, shorter (blue) wavelengths scatter more easily than longer (red) wavelengths. This is why the sky looks blue; you are seeing scattered blue light. This is also why sunsets are red. Because the Sun is close to the horizon, the Sun's rays pass through more atmosphere than normal to reach your eye. Much of the blue light has been scattered out, leaving the red light in a sunset.

Absorption

Different molecules absorb different wavelengths of radiation. For example, O_2 and O_3 absorb almost all wavelengths shorter than 300 nanometers. Water (H_2O) absorbs many wavelengths above 700 nm. When a molecule absorbs a photon, it increases the energy of the molecule. This heats the atmosphere, but the atmosphere also cools by emitting radiation, as discussed below.

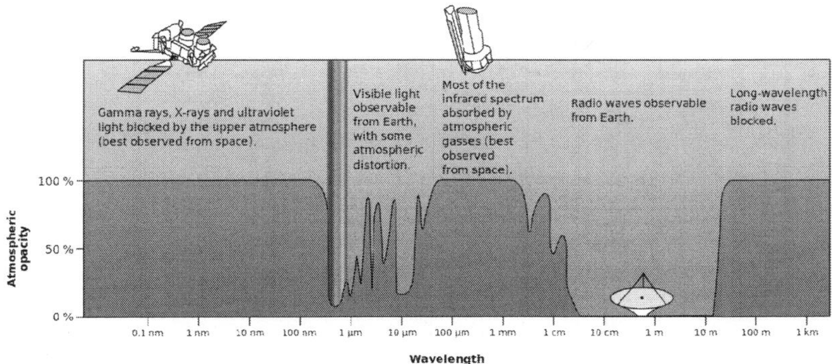

Figure: *Rough plot of Earth's atmospheric transmittance (or opacity) to various wavelengths of electromagnetic radiation, including visible light.*

The combined absorption spectra of the gases in the atmosphere leave "windows" of low opacity, allowing the transmission of only certain bands of light. The optical window runs from around 300 nm (ultraviolet-C) up into the range humans can see, the visible spectrum (commonly called light), at roughly 400–700 nm and continues to the infrared to around 1100 nm. There are also infrared and radio windows that transmit some infrared and radio waves at longer wavelengths. For example, the radio window runs from about one centimeter to about eleven-meter waves.

Emission

Emission is the opposite of absorption, it is when an object emits radiation. Objects tend to emit amounts and wavelengths of radiation depending on their "black body" emission curves, therefore hotter objects tend to emit more radiation, with shorter wavelengths. Colder objects emit less radiation, with longer wavelengths. For example, the Sun is approximately 6,000 K (5,730 °C; 10,340 °F), its radiation peaks near 500 nm, and is visible to the human eye. Earth is approximately 290 K (17 °C; 62 °F), so its radiation peaks near 10,000 nm, and is much too long to be visible to humans.

Because of its temperature, the atmosphere emits infrared radiation. For example, on clear nights Earth's surface cools down faster than on cloudy nights. This is because clouds (H_2O) are strong absorbers and emitters of infrared radiation. This is also why it becomes colder at night at higher elevations.

The greenhouse effect is directly related to this absorption and emission effect. Some gases in the atmosphere absorb and emit infrared radiation, but do not interact with sunlight in the visible spectrum. Common examples of these are CO_2 and H_2O.

Refractive Index

The refractive index of air is close to, but just greater than 1. Systematic variations in refractive index can lead to the bending of light rays over long optical paths. One example is that, under some circumstances, observers onboard ships can see other vessels just over the horizon because light is refracted in the same direction as the curvature of Earth's surface.

The refractive index of air depends on temperature, giving rise to refraction effects when the temperature gradient is large. An example of such effects is the mirage.

Atmospheric Circulation

Atmospheric circulation is the large-scale movement of air, and the means (together with the smaller ocean circulation) by which thermal energy is distributed on the surface of the Earth.

The large-scale structure of the atmospheric circulation varies from year to year, but the basic climatological structure remains fairly constant. Individual weather systems – mid-latitude depressions, or tropical convective cells – occur "randomly", and it is accepted that weather cannot be predicted beyond a fairly short limit: perhaps a month in theory, or (currently) about ten days in practice. Nonetheless, as the climate is the average of these systems and patterns – where and when they tend to occur again and again – it is stable over longer periods of time.

As a rule, the "cells" of Earth's atmosphere shift polewards in warmer climates (e.g. interglacials compared to glacials), but remain largely constant even due to continental drift; they are, fundamentally, a property of the Earth's size, rotation rate, heating and atmospheric depth, all of which change little. However, a tectonic uplift can significantly alter their major elements, for example, the jet stream, and plate tectonics may shift ocean currents. In the extremely hot climates of the Mesozoic, indications of a third desert belt at the Equator has been found; it was perhaps caused by convection. But even then, the overall latitudinal pattern of Earth's climate was not much different from the one today.

Latitudinal Circulation Features

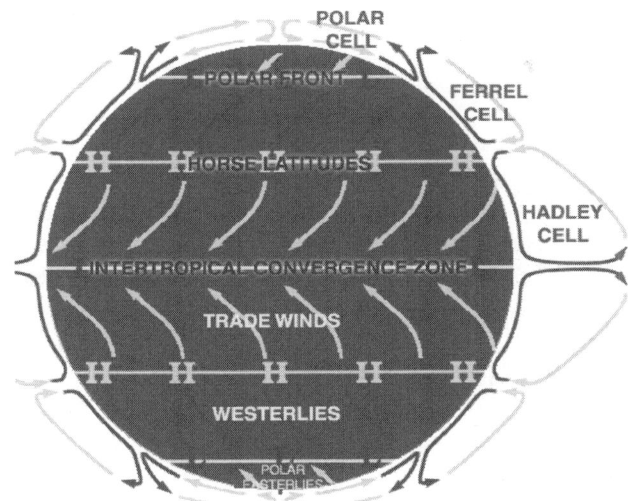

Figure: *An idealised view of three large circulation cells.*

The wind belts girdling the planet are organised into three cells: the Hadley cell, the Ferrel cell, and the Polar cell. Contrary to the impression given in the simplified diagram, the vast bulk of the vertical motion occurs in the Hadley cell; the explanations of the other two cells are complex. Note that there is one discrete Hadley cell that may split, shift and merge in a complicated process over time. Low and high pressures on earth's surface are balanced by opposite relative pressures in the upper troposphere.

Hadley Cell

The Hadley cell mechanism is well understood. The atmospheric circulation pattern that George Hadley described to provide an explanation for the trade winds matches observations very well. It is a closed circulation loop, which begins at the equator with warm, moist air lifted aloft in equatorial low-pressure areas (the Intertropical Convergence Zone, ITCZ) to the tropopause and carried poleward. At about 30°N/S latitude, it descends in a high-pressure area. Some of the descending air travels equatorially along the surface, closing the loop of the Hadley cell and creating the Trade Winds.

Though the Hadley cell is described as lying on the equator, it is more accurate to describe it as following the sun's zenith point, or what is termed the "thermal equator," which undergoes a semiannual north-south migration. The Hadley system provides an example of a thermally direct circulation. The thermodynamic efficiency and power of the Hadley system, considered as a heat engine, is estimated as 2.6% and 200 TW.

Polar Cell

The Polar cell is likewise a simple system. Though cool and dry relative to equatorial air, air masses at the 60th parallel are still sufficiently warm and moist to undergo convection and drive a thermal loop. Air circulates within the troposphere, limited vertically by the tropopause at about 8 km. Warm air rises at lower latitudes and moves poleward through the upper troposphere at both the north and south poles. When the air reaches the polar areas, it has cooled considerably, and descends as a cold, dry high-pressure area, moving away from the pole along the surface but veering westward as a result of the Coriolis effect to produce the Polar easterlies.

The outflow from the cell creates harmonic waves in the atmosphere known as Rossby waves. These ultra-long waves play an important role in determining the path of the jet stream, which travels within the transitional zone between the tropopause and the Ferrel cell. By acting as a heat sink, the Polar cell also balances the Hadley cell in the Earth's energy equation.

The Hadley cell and the Polar cell are similar in that they are thermally direct; in other words, they exist as a direct consequence of surface temperatures; their thermal characteristics override the effects of weather in their domain. The sheer volume of energy the Hadley cell transports, and the depth of the heat sink that is the Polar cell, ensures that the effects of transient weather phenomena are not only not felt by the system as a whole, but — except under unusual circumstances — are not even permitted to form. The endless chain of passing highs and lows which is part of everyday life for mid-latitude dwellers is unknown above the 60th and below the 30th parallels. There are some notable exceptions to this rule. In Europe, unstable weather extends to at least 70° north.

These atmospheric features are also stable, so even though they may strengthen or weaken regionally or over time, they do not vanish entirely.

The Polar cell, orography and Katabatic winds in Antarctica, can create very cold conditions at the surface, for instance the coldest temperature recorded on Earth: -89.2 °C at Vostok Station in Antarctica, measured 1983.

Ferrel Cell

The Ferrel cell, theorized by William Ferrel (1817–1891), is a secondary circulation feature, dependent for its existence upon the Hadley cell and the Polar cell. It behaves much as an atmospheric ball bearing between the Hadley cell and the Polar cell, and comes about as a result of the eddy circulations (the high- and low-pressure areas) of the mid-latitudes. For this reason it is sometimes known as the "zone of mixing." At its southern extent (in the Northern hemisphere), it overrides the Hadley cell, and at its northern extent, it overrides the Polar cell. Just as the Trade Winds can be found below the Hadley cell, the Westerlies can be found beneath the Ferrel cell. Thus, strong high-pressure areas which divert the prevailing westerlies, such as a Siberian high (which could be considered an extension of the Arctic high), could be said to override the Ferrel cell, making it discontinuous.

While the Hadley and Polar cells are truly closed loops, the Ferrel cell is not, and the telling point is in the Westerlies, which are more formally known as "the Prevailing Westerlies." While the Trade Winds and the Polar Easterlies have nothing over which to prevail, their parent circulation cells having taken care of any competition they might have to face, the Westerlies are at the mercy of passing weather systems. While upper-level winds are essentially westerly, surface winds can vary sharply and abruptly in direction. A low moving polewards or a high moving equator wards maintains or even accelerates a westerly flow; the local passage of a cold front may change that in a matter of minutes, and frequently does. A strong high moving polewards may bring easterly winds for days.

The base of the Ferrel cell is characterized by the movement of air masses, and the location of these air masses is influenced in part by the location of the jet stream, which acts as a collector for the air carried aloft by surface lows (a look at a weather map will show that surface lows follow the jet stream). The overall movement of surface air is from the 30th latitude to the 60th. However, the upper flow of the Ferrel cell is not well defined. This is in part because it is intermediary between the Hadley and Polar cells, with neither a strong heat source nor a strong cold sink to drive convection and, in part, because of the effects on the upper atmosphere of surface eddies, which act as destabilizing influences.

In contrast to the Hadley and Polar systems, the Ferrel system provides an example of a thermally indirect circulation. The Ferrel system acts as a heat pump with a coefficient of performance of 12.1, consuming kinetic energy at an approximate rate of 275 TW.

Longitudinal Circulation Features

While the Hadley, Ferrel, and Polar cells are major factors in global heat transport, they do not act alone. Disparities in temperature also drive a set of longitudinal circulation cells, and the overall atmospheric motion is known as the zonal overturning circulation.

Latitudinal circulation is the consequence of the fact that incident solar radiation per unit area is highest at the heat equator, and decreases as the latitude increases, reaching its minimum at the poles. Longitudinal circulation, on the other hand, comes about because water has a higher specific heat capacity than land and thereby absorbs and releases more heat, but the temperature changes less than land. Even at mesoscales (a horizontal range of 5 to several hundred kilometres), this effect is noticeable; it is what brings the sea breeze, air cooled by the water, ashore in the day, and carries the land breeze, air cooled by contact with the ground, out to sea during the night.

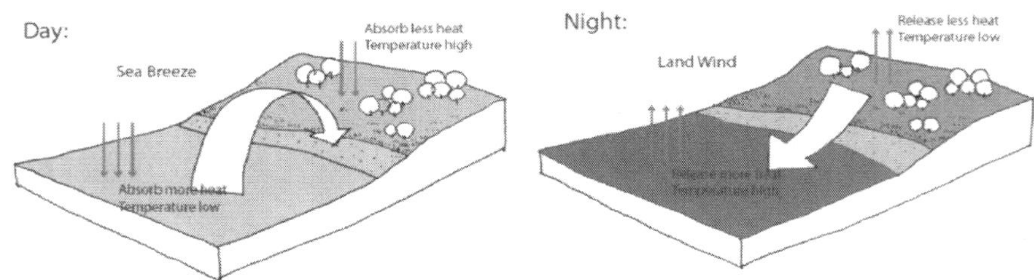

Figure: *Diurnal wind change in coastal area.*

On a larger scale, this effect ceases to be diurnal (daily), and instead is seasonal or even decadal in its effects. Warm air rises over the equatorial, continental, and western Pacific Ocean regions, flows eastward or westward, depending on its location, when it reaches the tropopause, and subsides in the Atlantic and Indian Oceans, and in the eastern Pacific.

The Pacific Ocean cell plays a particularly important role in Earth's weather. This entirely ocean-based cell comes about as the result of a marked difference in the surface temperatures of the western and eastern Pacific. Under ordinary circumstances, the western Pacific waters are warm and the eastern waters are cool. The process begins when strong convective activity over equatorial East Asia and subsiding cool air off South America's west coast creates a wind pattern which pushes Pacific water westward and piles it up in the western Pacific. (Water levels in the western Pacific are about 60 cm higher than in the eastern Pacific, a difference due entirely to the force of moving air.)

Walker Circulation

The Pacific cell is of such importance that it has been named the Walker circulation after Sir Gilbert Walker, an early-20th-century director of British observatories in India, who sought a means of predicting when the monsoon winds would fail. While he was never successful in doing so, his work led him to the discovery of an indisputable link between periodic pressure variations in the Indian Ocean and the Pacific, which he termed the "Southern Oscillation".

The movement of air in the Walker circulation affects the loops on either side. Under "normal" circumstances, the weather behaves as expected. But every few years, the winters become unusually warm or unusually cold, or the frequency of hurricanes increases or decreases, and the pattern sets in for an indeterminate period.

The behaviour of the Walker cell is the key to the riddle, and leads to an understanding of the El Nipo (more accurately, ENSO or El Niño – Southern Oscillation) phenomenon.

If convective activity slows in the Western Pacific for some reason (this reason is not currently known), the climate dominoes next to it begin to topple. First, the upper-level westerly winds fail. This cuts off the source of cool subsiding air, and therefore the surface Easterlies cease.

The consequence of this is twofold. In the eastern Pacific, warm water surges in from the west since there is no longer a surface wind to constrain it. This and the corresponding effects of the Southern Oscillation result in long-term unseasonable temperatures and precipitation patterns in North and South America, Australia, and Southeast Africa, and disruption of ocean currents.

Meanwhile in the Atlantic, high-level, fast-blowing Westerlies which would ordinarily be blocked by the Walker circulation and unable to reach such intensities, form. These winds tear apart the tops of nascent hurricanes and greatly diminish the number which are able to reach full strength.

El Niño – Southern Oscillation

El Niño and *La Niña* are two opposite surface temperature anomalies in the Southern Pacific, which heavily influence the weather on a large scale. In the case of El Niño. Warm water approaches the coasts of South America which results in blocking the upwelling of nutrient-rich deep water. This has serious impacts on the fish populations.

In the La Niña case, the convective cell over the western Pacific strengthens inordinately, resulting in colder than normal winters in North America, and a more robust cyclone season in South-East Asia and Eastern Australia. There is increased upwelling of deep cold ocean waters and more intense uprising of surface air near South America, resulting in increasing numbers of drought occurrences, although it is often argued that fishermen reap benefits from the more nutrient-filled eastern Pacific waters.

The neutral part of the cycle – the "normal" component – has been referred to humorously by some as "La Nada", which means "the nothing" in Spanish.

Evolution of Earth's Atmosphere

Earliest Atmosphere

The first atmosphere would have consisted of gases in the solar nebula, primarily hydrogen. In addition, there would probably have been simple hydrides such as those now found in gas giants like Jupiter and Saturn, notably water vapor, methane and ammonia. As the solar nebula dissipated, these gases would have escaped, partly driven off by the solar wind.

Second Atmosphere

The next atmosphere, consisting largely of nitrogen plus carbon dioxide and inert gases, was produced by outgassing from volcanism, supplemented by gases produced during the late heavy bombardment of Earth by huge asteroids. A major part of carbon dioxide emissions were soon dissolved in water and built up carbonate sediments.

Water-related sediments have been found dating from as early as 3.8 billion years ago. About 3.4 billion years ago, nitrogen was the major part of the then stable "second atmosphere". An influence of life has to be taken into account rather soon in the history of the atmosphere, because hints of early life forms are to be found as early as 3.5 billion years ago. How Earth at that time managed a climate warm enough for liquid water and life, if the early Sun put out 30% lower solar radiance than today, is a puzzle known as the "faint young Sun paradox".

The geological record however shows a continually relatively warm surface during the complete early temperature record of Earth with the exception of one cold glacial phase about 2.4 billion years ago. In the late Archaean eon an oxygen-containing atmosphere began to develop, apparently produced by photosynthesizing cyanobacteria which have been found as stromatolite fossils from 2.7 billion years ago. The early basic carbon isotopy (isotope ratio proportions) is very much in line with what is found today, suggesting that the fundamental features of the carbon cycle were established as early as 4 billion years ago.

Earth's dynamic oxygenation evolution is recorded in ancient sediments from the Republic of Gabon from between about 2,150 and 2,080 million years ago. These fluctuations in oxygenation were likely driven by the Lomagundi carbon isotope excursion.

Third Atmosphere

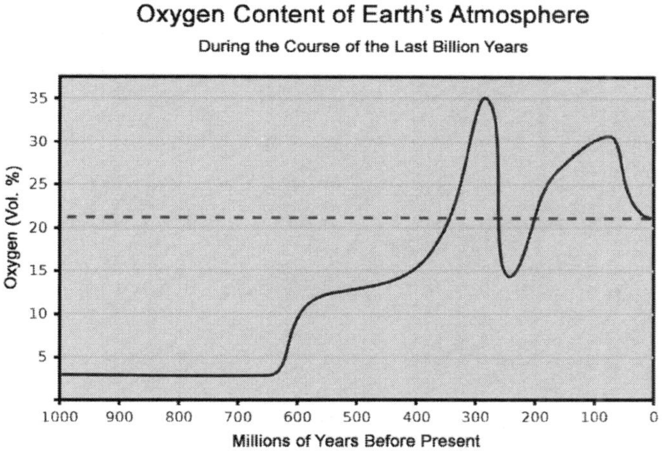

Figure: *Oxygen content of the atmosphere over the last billion years.*

The constant re-arrangement of continents by plate tectonics influences the long-term evolution of the atmosphere by transferring carbon dioxide to and from large continental carbonate stores. Free oxygen did not exist in the atmosphere until about 2.4 billion years ago during the Great Oxygenation Event and its appearance is indicated by the end of the banded iron formations. Before this time, any oxygen produced by photosynthesis was consumed by oxidation of reduced materials, notably iron. Molecules of free oxygen did not start to accumulate in the atmosphere until the rate of production of oxygen began to exceed the availability of reducing materials. This point signifies a shift from a reducing atmosphere to an oxidizing atmosphere. O_2 showed major variations until reaching a steady state of more than 15% by the end of the Precambrian. The following time span from 541 million years ago to the present day is the Phanerozoic eon, during the earliest period of which, the Cambrian, oxygen-requiring metazoan life forms began to appear.

The amount of oxygen in the atmosphere has fluctuated over the last 600 million years, reaching a peak of about 30% around 280 million years ago, significantly higher than today's 21%. Two main processes govern changes in the atmosphere: Plants use carbon dioxide from the atmosphere, releasing oxygen.

Breakdown of pyrite and volcanic eruptions release sulfur into the atmosphere, which oxidizes and hence reduces the amount of oxygen in the atmosphere. However, volcanic eruptions also release carbon dioxide, which plants can convert to oxygen. The exact cause of the variation of the amount of oxygen in the atmosphere is not known. Periods with much oxygen in the atmosphere are associated with rapid development of animals.

Today's atmosphere contains 21% oxygen, which is high enough for this rapid development of animals.

Currently, anthropogenic greenhouse gases are accumulating in the atmosphere, which is the main cause of global warming.

Near Space/ Upper Atmosphere Studies

Near space is the region of Earth's atmosphere that lies between 20 to 100 km (65,000 and 328,000 feet) above sea level, encompassing the stratosphere, mesosphere, and the lower thermosphere. It extends roughly from the Armstrong limit above which humans need a pressure suit to survive, up to the Kármán line where astrodynamics must take over from aerodynamics in order to achieve flight. Thus, near space is above where commercial airliners fly but below orbiting satellites.

The terms "near space" and "upper atmosphere" are generally considered synonymous. However, some sources distinguish between the two. Where such a distinction is made, only the layers closest to the Karman line are called near space, while only the remaining layers between the lower atmosphere and near space are called the upper atmosphere.

Exploration

Near space was first explored in the 1930s. The early flights flew to the edge of space without computers, spacesuits, and with only crude life support systems. Notable people who flew in near space were Jean Piccard and his wife Jeannette, on the nearcraft The Century of Progress. Later exploration was mainly carried out by unmanned nearcraft, although there have been skydiving attempts made from high altitude balloons.

Uses of Near Space

The area is of interest for military surveillance purposes, scientific study, as well as to commercial interests for communications, and tourism. Craft that fly in near space include high altitude balloons, non-rigid airships, rockoons, sounding rockets, and the Lockheed_U-2 aircraft. The region has been of interest to space travel. Early attempts used a craft known as a rockoon to reach extreme altitudes and orbit. These are still used today for sounding rockets.

There has been a resurgence of interest in near space to launch manned spacecraft by man. Groups like ARCASPACE, as well as the da Vinci Project are planning on launching manned suborbital space vehicles from high altitude balloons.

JP Aerospace has a proposal for a spaceport in near space, as part of their Airship to Orbit program.

Exosphere

The exosphere is a thin, atmosphere-like volume surrounding a planetary body where molecules are gravitationally bound to that body, but where the density is too low for them to behave as a gas by colliding with each other. In the case of bodies with substantial atmospheres, such as the Earth's atmosphere, the exosphere is the uppermost layer, where the atmosphere thins out and merges with interplanetary space. It is located directly above the thermosphere.

Several moons, such as Earth's moon and the Galilean satellites, have exospheres without a denser atmosphere underneath. Here molecules are ejected on parabolic trajectories until they collide with the surface. Authors differ as to whether such moons are considered to have atmospheres or not. Smaller bodies such as asteroids, in which the molecules emitted from the surface escape to space, are not considered to have exospheres.

The main gases within the Earth's are the lightest atmospheric gases, mainly hydrogen, with some helium, carbon dioxide, and atomic oxygen near the base of the exosphere. Since there is no clear boundary between outer space and the exosphere, the exosphere is sometimes considered a part of outer space.

Lower Boundary

The lower boundary of the exosphere is known as *exopause*; it is also called the *exobase*, as in Earth's atmosphere the atmospheric temperature becomes nearly a constant above this altitude. Before the term exobase was established the boundary was also called the *critical altitude* where barometric conditions no longer apply. The altitude of the exobase ranges from about 500 to 1,000 kilometres (310 to 620 mi) depending on solar activity.

The Exobase can be Defined in One of Two Ways

If we define the exobase as the height at which upward-travelling molecules experience one collision on average, then at this position the mean free path of a molecule is equal to one pressure scale height. This is shown in the following. Consider a volume of air, with horizontal area A and height equal to the mean free path l, at pressure p and temperature T. For an ideal gas, the number of molecules contained in it is:

$$n = \frac{pAl}{RT}$$

where R is the universal gas constant. From the requirement that each molecule travelling upward undergoes on average one collision, the pressure is:

$$p = \frac{m_A ng}{A}$$

where m_A is the mean molecular mass of the gas. Solving these two equations gives:

$$l = \frac{RT}{m_A g}$$

which is the equation for the pressure scale height. As the pressure scale height is almost equal to the density scale height of the primary constituent, and since the Knudsen number is the ratio of mean free path and typical density fluctuation scale, this means that the exobase lies in the region where $\mathrm{Kn}(h_{EB}) \simeq 1$.

The fluctuation in the height of the exobase is important because this provides atmospheric drag on satellites, eventually causing them to fall from orbit if no action is taken to maintain the orbit.

Upper Boundary

In principle, the exosphere covers all distances where particles are still gravitationally bound to Earth, i.e. particles still have ballistic orbits that will take them back towards Earth. The upper boundary of the exosphere can be defined as the distance at which the influence of solar radiation pressure on atomic hydrogen exceeds that of the Earth's gravitational pull. This happens at half the distance to the Moon

(190,000 kilometres (120,000 mi)). The exosphere observable from space as the geocorona is seen to extend to at least 10,000 kilometres (6,200 mi) from the surface of the Earth. The exosphere is a transitional zone between Earth's atmosphere and interplanetary space.

Upper-Atmospheric Models

Most climate models simulate a region of the Earth's atmosphere from the surface to the stratopause. There also exist numerical models which simulate the wind, temperature and composition of the Earth's tenuous upper atmosphere, from the mesosphere to the exosphere, including the ionosphere. This region is affected strongly by the 11 year Solar cycle through variations in solar UV/EUV/Xray radiation and solar wind leading to high latitude particle precipitation and aurora. It has been proposed that these phenomena may have an effect on the lower atmosphere, and should therefore be included in simulations of climate change. For this reason there has been a drive in recent years to create "whole atmosphere" models to investigate whether or not this is the case.

A jet stream perturbation model is employed by Weather Logistics UK, which simulates the diversion of the air streams in the upper atmosphere. North Atlantic air flow modelling is simulated by combining a monthly jet stream climatology input calculated at 20 to 30°W, with different blocking high patterns. The jet stream input is generated by thermal wind balance calculations at 316mbars (6 to 9 km aloft) in the mid-latitude range from 40 to 60°N. Long term blocking patterns are determined by the weather forecaster, who identifies the likely position and strength of North Atlantic Highs from synoptic charts, the North Atlantic Oscillation (NAO) and El Niño-Southern Oscillation (ENSO) patterns.

The model is based on the knowledge that low pressure systems at the surface are steered by the fast ribbons (jet streams) of air in the upper atmosphere. The jet stream - blocking interaction model simulation examines the sea surface temperature field using data from NOAA tracked along the ocean on a path to the British Isles. The principal theory suggests that long term weather patterns act on longer time scales, so large blocking patterns are thought to appear in a similar locations repeatedly over several months. With a good knowledge of blocking high patterns, the model performs with an impressive accuracy that is useful to the end user.

The modelling undertaken at Weather Logistics UK produces regional-seasonal predictions that are probabilistic in nature. Two different blocking sizes are used for the modelling, located at two different locations. The four possible blocking diversions are then ranked in an order, to be combined by logistic regression and generate the appropriate likelihoods of weather events on seasonal time-scales. The raw output consists of 22 different weather conditions for each season that are compared to the average atmospheric conditions.

A global warming bias and 1961–1990 climatology of regional British Isles temperatures are added to the anomaly value to produce a final temperature prediction. The seasonal weather forecasts at Weather Logistics UK comprise of several additional weather components (derivatives) including: precipitation anomlies, storm tracks, air flow trajectories, heating degree days for household utility bills, cooling degree days, heat wave and the snow day odds.

According to a report in New Scientist many researchers are in consensus that Rossby waves are acting against the jet stream's usual pattern and holding it in place. Upper atmospheric studies using National Oceanic and Atmospheric Administration (NOAA) data indicates that during July 2010 these upper air stream patterns were most frequently observed in the Northern Hemisphere. Examination of the climatology data over the same period of time indicates that these wild planetary wave meanderings are not a normal aspect of our regional climate patterns.

Meanwhile, ongoing research studies at the University of Reading show that unusual patterns in the polar jet stream are more common during a period of low activity in the solar cycle when the observed sun spot activity and their associated solar flares are at their minimum. The link between low solar activity and enhanced blocking patterns is associated with an increase in the prevalence of cold weather patterns during the European Winter.

Another possible explanation for the observed increase in blocking patterns is natural variability, through the chaotic character of the large-scale Ocean currents that flow across the surface of the tropical Pacific.

Kármán Line

The Kármán line, or Karman line, lies at an altitude of 100 kilometres (62 mi) above the Earth's sea level, and commonly represents the boundary between the Earth's atmosphere and outer space. This definition is accepted by the Fédération Aéronautique Internationale (FAI), which is an international standard setting and record-keeping body for aeronautics and astronautics.

The line is named after Theodore von Kármán (1881–1963), a Hungarian-American engineer and physicist. He was active primarily in aeronautics and astronautics. He was the first to calculate that around this altitude, the atmosphere becomes too thin to support aeronautical flight, because a vehicle at this altitude would have to travel faster than orbital velocity to derive sufficient aerodynamic lift to support itself (neglecting centrifugal force). There is an abrupt increase in atmospheric temperature and interaction with solar radiation just below the line, which places the line within the greater thermosphere.

Definition

An atmosphere does not abruptly end at any given height, but becomes progressively thinner with altitude. Also, depending on how the various layers that make up the space around the Earth are defined (and depending on whether these layers are considered part of the actual atmosphere), the definition of the edge of space could vary considerably: If one were to consider the thermosphere and exosphere part of the atmosphere and not of space, one might have to extend the boundary to space to at least 10,000 km (6,200 mi) above sea level. The Kármán line thus is an arbitrary definition based on the following considerations:

An aircraft only stays in the sky if it constantly travels forward relative to the air (airspeed is not dependent on speed relative to ground), so that the wings can generate lift. The thinner the air, the faster the plane must go to generate enough lift to stay up.

If the lift coefficient for a wing at a specified angle of attack is known (or estimated using a method such as thin-airfoil theory), then the lift produced for specific flow conditions can be determined using the following equation

$$L = \tfrac{1}{2}\rho v^2 A C_L$$

where

L is lift force

ρ is air density

v is speed relative to the air

A is wing area,

C_L is the lift coefficient at the desired angle of attack, Mach number, and Reynolds number.

Lift (L) generated is directly proportional to the air density (ρ). All other factors remaining unchanged, true airspeed (v) must increase to compensate for less air density (ρ) at higher altitudes.

An orbiting spacecraft only stays in the sky if the centrifugal component of its movement around the Earth is enough to balance the downward pull of gravity. If it goes slower, the pull of gravity gradually makes its altitude decrease. The required speed is called *orbital velocity,* and it varies with the height of the orbit. For the International Space Station, or a space shuttle in low Earth orbit, the orbital velocity is about 27,000 km per hour (17,000 miles per hour).

For an airplane flying higher and higher, the increasingly thin air provides less and less lift, requiring increasingly higher speed to create enough lift to hold the airplane up. It eventually reaches an altitude where it must fly so fast to generate lift that it reaches orbital velocity. The Kármán line is the altitude where the speed necessary to aerodynamically support the airplane's full weight equals orbital velocity (assuming wing loading of a typical airplane). In practice, supporting full weight wouldn't be necessary to maintain altitude because the curvature of the Earth adds centrifugal lift as the airplane reaches orbital speed. However, the Karman line definition ignores this effect because orbital velocity is implicitly sufficient to maintain any altitude regardless of atmospheric density. The Karman line is therefore the highest altitude at which orbital speed provides sufficient aerodynamic lift to fly in a straight line that doesn't follow the curvature of the Earth's surface. When studying aeronautics and astronautics in the 1950s, Kármán calculated that above an altitude of roughly 100 km (62 mi), a vehicle would have to fly faster than orbital velocity to derive sufficient aerodynamic lift from the atmosphere to support itself. At this altitude, the air density is about 1/2200000 the density on the surface. At the Karman line, the air density ρ is such that

$$L = \frac{1}{2}\rho v_0^2 A C_L = mg$$

where

v_0 is orbital velocity

m is mass of the aircraft

g is acceleration due to gravity.

Although the calculated altitude was not exactly 100 km, Kármán proposed that 100 km be the designated boundary to space, because the round number is more memorable, and the calculated altitude varies minutely as certain parameters are varied. An international committee recommended the 100 km line to the FAI, and upon adoption, it became widely accepted as the boundary to space for many purposes. However, there is still no international legal definition of the demarcation between a country's air space and outer space.

Another hurdle to strictly defining the boundary to space is the dynamic nature of Earth's atmosphere. For example, at an altitude of 1,000 km (620 mi), the atmosphere's density can vary by a factor of five, depending on the time of day, time of year, AP magnetic index, and recent solar flux.

The FAI uses the Kármán line to define the boundary between aeronautics and astronautics:

- Aeronautics — For FAI purposes, aerial activity, including all air sports, within 100 kilometres of Earth's surface.
- Astronautics — For FAI purposes, activity more than 100 kilometres above Earth's surface.

Interpretations of the Definition

Some people (including the FAI in some of their publications) also use the expression "edge of space" to refer to a region below the conventional 100 km boundary to space, which is often meant to include substantially lower regions as well. Thus, certain balloon or airplane flights might be described as "reaching the edge of space". In such statements, "reaching the edge of space" merely refers to going higher than average aeronautical vehicles commonly would.

Alternatives to the Definition

The U.S. Air Force definition of an astronaut is a person who has flown more than 50 miles (~80 km) above mean sea level, approximately the line between the mesosphere and the thermosphere. NASA uses the FAI's 100-kilometer figure. The United States does not officially define a *boundary of space*. In 2005, three veteran NASA X-15 pilots (John B. McKay, Bill Dana and Joseph Albert Walker) were retroactively (two posthumously) awarded their astronaut wings, as they had flown between 90 and 108 km in the 1960s, but at the time had not been recognised as astronauts.

International law defines the lower boundary of space as the lowest perigee attainable by an orbiting space vehicle, but does not specify an altitude. Due to atmospheric drag, the lowest altitude at which an object in a circular orbit can complete at least one full revolution without propulsion is approximately 150 km (90 mi), whereas an object can maintain an elliptical orbit with perigee as low as about 130 km (80 mi) with propulsion. Above altitudes of approximately 160 km (100 mi) the sky is completely black.

Chapter 4

Pointings Energy Conservation Theorem

In electrodynamics, Pointing's theorem is a statement of conservation of energy for the electromagnetic field, in the form of a partial differential equation, due to the British physicist John Henry Pointing. Pointing's theorem is analogous to the work-energy theorem in classical mechanics, and mathematically similar to the continuity equation, because it relates the energy stored in the electromagnetic field to the work done on a charge distribution (i.e. an electrically charged object), through energy flux.

Statement

General

In words, the theorem is an energy balance:

The rate of energy transfer (per unit volume) from a region of space equals the rate of work done on a charge distribution plus the energy flux leaving that region.

Mathematically, this is Summarised in Differential Form as

$$-\frac{\partial u}{\partial t} = \nabla \cdot \mathrm{S} + \mathrm{J}_f \cdot \mathrm{E}$$

where $\nabla \cdot \mathrm{S}$ is the divergence of the Pointing vector (energy flow) and $\mathrm{J} \cdot \mathrm{E}$ is the rate at which the fields do work on a charged object (J is the *free* current density corresponding to the motion of charge, E is the electric field, and $\cdot$ is the dot product). The energy density u is given by:

$$u = \frac{1}{2}(E \cdot D + B \cdot H)$$

in which D is the electric displacement field, B is the magnetic flux density and H the magnetic field strength, ε_0 is the electric constant and μ_0 is the magnetic constant.

Since the charges are free to move, and the D and H fields bypass any bound charges and currents in the charge distribution (by their definition), J is the *free* current density, *not* the *total*.

Using the divergence theorem, Pointing's theorem can be rewritten in integral form:

$$-\frac{\partial}{\partial t}\int_V u dV = S \cdot dA + \int_V J \cdot E dV$$

where ∂V is the boundary of a volume V. The shape of the volume is arbitrary but fixed for the calculation.

Electrical Engineering

In electrical engineering context the theorem is usually written with the energy density term u expanded in the following way, which resembles the continuity equation:

$$\nabla \cdot S + \epsilon_0 E \cdot \frac{\partial E}{\partial t} + \frac{B}{\mu_0} \cdot \frac{\partial B}{\partial t} + J \cdot E = 0,$$

where

- $\epsilon_0 E \cdot \frac{\partial E}{\partial t}$ is the density of reactive power driving the build-up of electric field,
- $\frac{B}{\mu_0} \cdot \frac{\partial B}{\partial t}$ is the density of reactive power driving the build-up of magnetic field, and
- $J \cdot E$ is the density of Electric power dissipated by the Lorentz force acting on charge carriers.

Derivation

While conservation of energy and the Lorentz force law can derive the general form of the theorem, Maxwell's equations are additionally required to derive the expression for the Pointing vector and hence complete the statement.

Pointing's Theorem

Considering the statement in words above - there are three elements to the theorem, which involve writing energy transfer (per unit time) as volume integrals:

1. Since u is the energy density, integrating over the volume of the region gives the total energy U stored in the region, then taking the (partial) time derivative gives the rate of change of energy:

$$U = \int_V u dV \rightarrow \frac{\partial U}{\partial t} = \frac{\partial}{\partial t}\int_V u dV = \int_V \frac{\partial u}{\partial t} dV.$$

2. The energy flux leaving the region is the surface integral of the Pointing vector, and using the divergence theorem this can be written as a volume integral:

$$\oiint_{\partial V} \mathbf{S}\cdot d\mathbf{A} = \int_V \nabla\cdot\mathbf{S} dV.$$

3. The Lorentz force density f on a charge distribution, integrated over the volume to get the total force F, is

$$\mathrm{f} = \rho\mathrm{E} + \mathrm{J}\times\mathrm{B} \rightarrow \int_V \mathrm{f} dV = \mathrm{F} = \int_V (\rho\mathrm{E} + \mathrm{J}\times\mathrm{B}) dV,$$

where ρ is the charge density of the distribution and v its velocity. Since $\mathbf{J} = \rho\mathbf{v}$, the rate of work done by the force is

$$\mathrm{F}\cdot\frac{d\mathrm{r}}{dt} = \mathrm{F}\cdot\mathrm{v} = \int_V (\rho\mathrm{E}\cdot\mathrm{v} + \rho\mathrm{v}\times\mathrm{B}\cdot\mathrm{v}) dV \rightarrow \mathrm{F}\cdot\mathrm{v} = \int_V \mathrm{E}\cdot\mathrm{J} dV.$$

So by conservation of energy, the balance equation for the energy flow per unit time is the integral form of the theorem:

$$-\int_V \frac{\partial u}{\partial t} dV = \int_V \nabla\cdot\mathbf{S} dV + \int_V \mathbf{J}\cdot\mathbf{E} dV,$$

and since the volume V is arbitrary, this is true for all volumes, implying

$$-\frac{\partial u}{\partial t} = \nabla\cdot\mathbf{S} + \mathbf{J}\cdot\mathbf{E},$$

which is Pointing's theorem in differential form.

Pointing Vector

Pointing vector represents the directional energy flux density (the rate of energy transfer per unit area, in units of watts per square metre ($\mathrm{W\cdot m^{-2}}$)) of an electromagnetic field. It is named after its inventor John

Henry Pointing. Oliver Heaviside and Nikolay Umov independently co-invented the Pointing vector.

In Pointing's original paper and in many textbooks, it is usually denoted by S or N, and defined as:

$$S = E \times H,$$

which is often called the Abraham form; where E is the electric field and H the magnetic field. (All bold letters represent vectors.)

Occasionally an alternative definition in terms of electric field E and the magnetic flux density B is used. It is even possible to combine the displacement field D with the magnetic flux density B to get the Minkowski form of the Pointing vector, or use D and H to construct another. The choice has been controversial: Pfeifer et al. summarize and to a certain extent resolve the century-long dispute between proponents of the Abraham and Minkowski forms.

The Pointing vector represents the particular case of an energy flux vector for electromagnetic energy. However, any type of energy has its direction of movement in space, as well as its density, so energy flux vectors can be defined for other types of energy as well, e.g., for mechanical energy. The Umov–Pointing vector discovered by Nikolay Umov in 1874 describes energy flux in liquid and elastic media in a completely generalized view.

From the theorem, the actual form of the Pointing vector S can be found. The time derivative of the energy density (using the product rule for vector dot products) is

$$\frac{\partial u}{\partial t} = \frac{1}{2}\left(E \cdot \frac{\partial D}{\partial t} + D \cdot \frac{\partial E}{\partial t} + H \cdot \frac{\partial B}{\partial t} + B \cdot \frac{\partial H}{\partial t}\right) = E \cdot \frac{\partial D}{\partial t} + H \cdot \frac{\partial B}{\partial t},$$

using the constitutive relations

$$D = \epsilon_0 E, \quad B = \mu_0 H.$$

The partial time derivatives suggest using two of Maxwell's Equations. Taking the dot product of the Maxwell–Faraday equation with H:

$$\frac{\partial B}{\partial t} = -\nabla \times E \rightarrow H \cdot \frac{\partial B}{\partial t} = -H \cdot \nabla \times E,$$

next taking the dot product of the Maxwell–Ampère equation with E:

$$\frac{\partial D}{\partial t} + J = \nabla \times H \rightarrow E \cdot \frac{\partial D}{\partial t} + E \cdot J = E \cdot \nabla \times H.$$

Collecting the results so far gives:

$$
\begin{aligned}
-\nabla\cdot S &= \frac{\partial u}{\partial t} + J\cdot E \\
&= \left(H\cdot\frac{\partial B}{\partial t} + E\cdot\frac{\partial D}{\partial t} \right) + J\cdot E \\
&= E\cdot\nabla\times H - H\cdot\nabla\times E,
\end{aligned}
$$

then, using the vector calculus identity:

$$\nabla\cdot E\times H = H\cdot\nabla\times E - E\cdot\nabla\times H,$$

gives an expression for the Pointing vector:

$$S = E\times H,$$

which physically means the energy transfer due to time-varying electric and magnetic fields is perpendicular to the fields.

Alternative Forms

It is possible to derive alternative versions of Pointing's theorem. Instead of the flux vector E × B as above, it is possible to follow the same style of derivation, but instead choose the Abraham form E × H, the Minkowski form D × B, or perhaps D × H. Each choice represents the response of the propagation medium in its own way: the E × B form above has the property that the response happens only due to electric currents, while the D × H form uses only (fictitious) magnetic monopole currents.

The other two forms (Abraham and Minkowski) use complementary combinations of electric and magnetic currents to represent the polarization and magnetization responses of the medium.

Generalization

The *mechanical* energy counterpart of the above theorem for the *electromagnetic* energy continuity equation is

$$\frac{\partial}{\partial t}u_m(\mathbf{r},t) + \nabla\cdot\mathbf{S}_m(\mathbf{r},t) = \mathbf{J}(\mathbf{r},t)\cdot\mathbf{E}(\mathbf{r},t),$$

where u_m is the (mechanical) kinetic energy density in the system. It can be described as the sum of kinetic energies of particles α (e.g., electrons in a wire), whose trajectory is given by $r_\alpha(t)$:

$$u_m(\mathbf{r},t) = \sum_\alpha \frac{m_\alpha}{2}\dot{r}_\alpha^2\delta(\mathbf{r} - \mathbf{r}_\alpha(t)),$$

where S_m is the flux of their energies, or a "mechanical Pointing vector":

$$\mathbf{S}_m(\mathbf{r},t) = \sum_\alpha \frac{m_\alpha}{2} \dot{r}_\alpha^2 \, \mathbf{r}_\alpha \, \delta(\mathbf{r} - \mathbf{r}_\alpha(t)).$$

Both can be combined via the Lorentz force, which the electromagnetic fields exert on the moving charged particles, to the following energy continuity equation or energy conservation law:

$$\frac{\partial}{\partial t}(u_e + u_m) + \nabla \cdot (\mathbf{S}_e + \mathbf{S}_m) = 0,$$

covering both types of energy and the conversion of one into the other.

Magnetopause

The magnetopause is the abrupt boundary between a magnetosphere and the surrounding plasma. For planetary science, the magnetopause is the boundary between the planet's magnetic field and the solar wind. The location of the magnetopause is determined by the balance between the pressure of the dynamic planetary magnetic field and the dynamic pressure of the solar wind. As the solar wind pressure increases and decreases, the magnetopause moves inward and outward in response. Waves (ripples and flapping motion) along the magnetopause move in the direction of the solar wind flow in response to small scale variations in the solar wind pressure and to Kelvin-Helmholtz instability.

The solar wind is supersonic and passes through a bow shock where the direction of flow is changed so that most of the solar wind plasma is deflected to either side of the magnetopause, much like water is deflected before the bow of a ship. The zone of shocked solar wind plasma is the magnetosheath. At Earth and all the other planets with intrinsic magnetic fields, some solar wind plasma succeeds in entering and becoming trapped within the magnetosphere. At Earth, the solar wind plasma which enters the magnetosphere forms the plasma sheet. The amount of solar wind plasma and energy that enters the magnetosphere is regulated by the orientation of the interplanetary magnetic field, which is embedded in the solar wind.

The Sun and other stars with magnetic fields and stellar winds have a solar magnetopause or heliopause where the stellar environment is bounded by the interstellar environment.

Characteristics

Prior to the age of space exploration, interplanetary space was considered to be a vacuum. The coincidence of the Carrington super flare and the super geomagnetic event of 1859 was evidence that plasma was ejected from the Sun during a flare event. Chapman and Ferraro proposed that a plasma was emitted by the Sun in a burst as part of a flare event which disturbed the planet's magnetic field in a manner known as a geomagnetic storm.

The collision frequency of particles in the plasma in the interplanetary medium is very low and the electrical conductivity is so high that it could be approximated to an infinite conductor.

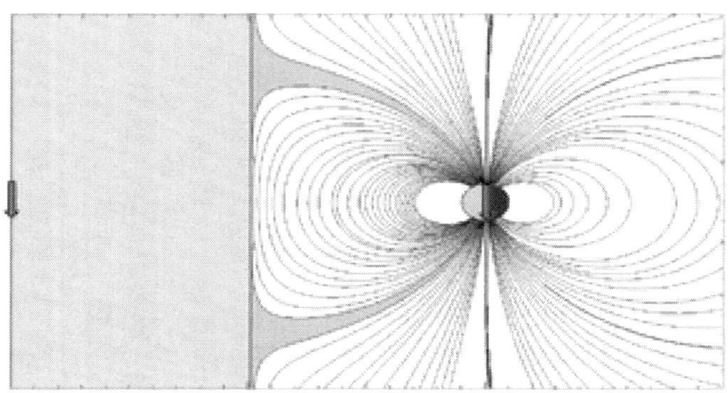

Figure: *Schematic representation of a planetary dipole magnetic field in a vacuum (right side) deformed by a region of plasma with infinite conductivity. The Sun is to the left. The configuration is equivalent to an image dipole (green arrow) being placed at twice the distance from the planetary dipole to the interaction boundary.*

A magnetic field in a vacuum cannot penetrate a volume with infinite conductivity. Chapman and Bartels (1940) illustrated this concept by postulating a plate with infinite conductivity placed on the dayside of a planet's dipole as shown in the schematic. The field lines on the dayside are bent.

At low latitudes, the magnetic field lines are pushed inward. At high latitudes, the magnetic field lines are push backwards and over the polar regions. The boundary between the region dominated by the planet's magnetic field (i.e., the magnetosphere) and the plasma in the interplanetary medium is the magnetopause.

The configuration equivalent to a flat, infinitely conductive plate is achieved by placing an image dipole (green arrow at left of schematic) at twice the distance from the planet's dipole to the magnetopause along the planet-Sun line. Since the solar wind is continuously flowing outward, the magnetopause above, below and to the sides of the planet are swept backward into the geomagnetic tail as shown in the artist's concept.

The region (shown in pink in the schematic) which separates field lines from the planet which are pushed inward from those which are pushed backward over the poles is an area of weak magnetic field or day-side cusp.

Solar wind particles can enter the planet's magnetosphere through the cusp region. Because the solar wind exists at all times and not just times of solar flares, the magnetopause is a permanent feature of the space near any planet with a magnetic field.

The magnetic field lines of the planet's magnetic field are not stationary. They are continuously joining or merging with magnetic field lines of the interplanetary magnetic field. The joined field lines are swept back over the poles into the planetary magnetic tail. In the tail, the field lines from the planet's magnetic field are re-joined and start moving toward night-side of the planet. The physics of this process was first explain by Dungey (1961) .

If one assumed that magnetopause was just a boundary between a magnetic field in a vacuum and a plasma with a weak magnetic field embedded in it, then the magnetopause would be defined by electrons and ions penetrating one gyroradius into the magnetic field domain. Since the gyro-motion of electrons and ions is in opposite directions, an electric current flows along the boundary. The actual magnetopause is much more complex.

Estimating the Standoff Distance to the Magnetopause

If the pressure from particles within the magnetosphere is neglected, it is possible to estimate the distance to the part of the magnetosphere that faces the Sun. The condition governing this position is that the dynamic ram pressure from the solar wind is equal to the magnetic pressure from the Earth's magnetic field.

$(\rho v^2)_{sw} \approx \left(\dfrac{4B(r)^2}{2\mu_0}\right)_m$ where ρ and v are the density and velocity of the solar wind, and

$B(r)$ is the Magnetic field strength of the planet in SI units (B in T, μ_0 in H/m)

Since the dipole magnetic field strength varies with distance as $1/r^3$ the magnetic field strength can be written as $B(r) = B_0 / r^3$.

$$\rho v^2 \approx \frac{2B_0^2}{r^6 \mu_0}.$$

Overview of the Solar System magnetopauses					
Planet	***Number***	***Magnetic moment***	***Magnetopause distance***	***Observed size of the magnetosphere***	***variance of magnetosphere***
Mercury	1	0.0004	1.5	1.4	0
Venus	2	0	0	0	0
Earth	3	1	10	10	2
Mars	4	0	0	0	0
Jupiter	5	20000	42	75	25
Saturn	6	600	19	19	3
Uranus	7	50	25	18	0
Neptune	8	25	24	24.5	1.5

Solving this equation for r leads to an estimate of the distance

$$r \approx \sqrt[6]{\frac{2B_0^2}{\mu_0 \rho v^2}}$$

The distance from Earth to the subsolar magnetopause varies over time due to solar activity, but typical distances range from 6 - 15 $R_{\oplus}$. Empirical models using real-time solar wind data can provide a real-time estimate of the magnetopause location. A bow shock stands upstream from the magnetopause. It serves to decelerate and deflect the solar wind flow before it reaches the magnetopause

Solar System Magnetopauses

Research on the magnetopause is conducted using the LMN coordinate system (which is set of axes like XYZ). N points normal to the magnetopause outward to the magnetosheath, L lies along the projection of the dipole axis onto the magnetopause (positive northward), and M completes the triad by pointing dawnward.

Venus and Mars do not have a planetary magnetic field and do not have a magnetopause. The solar wind interacts with the planet's atmosphere and a void is created behind the planet. In the case of the Earth's moon and other bodies without a magnetic field or atmosphere, the body's surface interacts with the solar wind and a void is created behind the body.

Kelvin–Helmholtz Instability

The Kelvin–Helmholtz instability (after Lord Kelvin and Hermann von Helmholtz) can occur when there is velocity shear in a single continuous fluid, or where there is a velocity difference across the interface between two fluids. An example is wind blowing over water: The instability manifests in waves on the water surface. More generally, clouds, the ocean, Saturn's bands, Jupiter's Red Spot, and the sun's corona show this instability.

The theory predicts the onset of instability and transition to turbulent flow in fluids of different densities moving at various speeds. Helmholtz studied the dynamics of two fluids of different densities when a small disturbance, such as a wave, was introduced at the boundary connecting the fluids.

For some short enough wavelengths, if surface tension is ignored, two fluids in parallel motion with different velocities and densities yield an interface that is unstable for all speeds. Surface tension stabilises the short wavelength instability however, and theory predicts stability until a velocity threshold is reached. The theory with surface tension included broadly predicts the onset of wave formation in the important case of wind over water. In gravity, for a continuously varying distribution of density and velocity (with the lighter layers uppermost, so that the fluid is RT-stable), the dynamics of the KH instability is described by the Taylor–Goldstein equation and its onset is given by a Richardson number, Ri. Typically the layer is unstable for Ri<0.25. These effects are common in cloud layers. The study of this instability is applicable in plasma physics, for example in inertial confinement fusion and the plasma–beryllium interface.

Numerically, the KH instability is simulated in a temporal or a spatial approach. In the temporal approach, experimenters consider the flow in a periodic (cyclic) box "moving" at mean speed (absolute instability). In the spatial approach, experimenters simulate a lab experiment with natural inlet and outlet conditions (convective instability).

Magnetic Sail

A magnetic sail or magsail is a proposed method of spacecraft propulsion which would use a static magnetic field to deflect charged particles radiated by the Sun as a plasma wind, and thus impart momentum to accelerate the spacecraft. A magnetic sail could also thrust directly against planetary and solar magnetospheres.

The magnetic sail was invented by Dana Andrews and Robert Zubrin working in collaboration in 1988. At that time, Andrews was working on a concept to use a magnetic scoop to gather ions to provide propellant for a nuclear electric ion drive spacecraft, allowing the craft to operate in the same manner of a Bussard ramjet, but without the need for a proton-proton fusion propulsion drive. He asked Zubrin to help him compute the drag that the magnetic scoop would create against the interplanetary medium. Zubrin agreed, but found that the drag created by the scoop would be much greater than the thrust created by the ion drive. He therefore proposed that the ion drive component of the system be dropped, and the device simply used as a sail. Andrews agreed, and the magsail was born. The two then proceeded to elaborate their analysis of the magsail for interplanetary, interstellar, and planetary orbital propulsion in a series of papers published from 1988 through the 1990s.

Principles of Operation and Design

The *magsail* operates by creating drag against the local medium (planet's magnetic field, solar wind, or interstellar winds), thereby allowing a spacecraft accelerated to very high velocities by other means, such as a fusion rocket or laser pushed lightsail, to slow down - even from relativistic velocities - without requiring the use of onboard propellant. It can thus reduce the delta-V propulsion required for an interstellar mission by a factor of two. This capability is the most unique feature of the magsail, and perhaps the most significant in the long term.

In typical magnetic sail designs, the magnetic field is generated by a loop of superconducting wire. Because loops of current-carrying conductors tend to be forced outwards towards a circular shape by their own magnetic field, the sail could be deployed simply by unspooling the conductor and applying a current through it.

Solar Wind Example

The solar wind is a continuous stream of plasma that flows outwards from the Sun: near the Earth's orbit, it contains several million protons and electrons per cubic meter and flows at 400 to 600 km/s (250 to 370 mi/s). The magnetic sail introduces a magnetic field into this plasma flow which can deflect the particles from their original trajectory: the momentum of the particles is then transferred to the sail, leading to a thrust on the sail. One advantage of magnetic or solar sails over (chemical or ion) reaction thrusters is that no reaction mass is depleted or carried in the craft.

For a sail in the solar wind one AU away from the Sun, the field strength required to resist the dynamic pressure of the solar wind is 50 nT. Zubrin's proposed magnetic sail design would create a bubble of space of 100 km in diameter (62 mi) where solar-wind ions are substantially deflected using a hoop 50 km (31 mi) in radius. The minimum mass of such a coil is constrained by material strength limitations at roughly 40 tonnes (44 tons) and it would generate 70 N (16 lb_f) of thrust, giving a mass/thrust ratio of 600 kg/N. If operated within the solar system, high temperature superconducting wire would be required to make the magsail practical. If operated in interstellar space conventional superconductors would be adequate.

The operation of magnetic sails using plasma wind is analogous to the operation of solar sails using the radiation pressure of photons emitted by the Sun. Although solar wind particles have rest mass and photons do not, sunlight has thousands of times more momentum than the solar wind. Therefore, a magnetic sail must deflect a proportionally larger area of the solar wind than a comparable solar sail to generate the same amount of thrust. However, it need not be as massive as a solar sail because the solar wind is deflected by a magnetic field instead of a large physical sail. Conventional materials for solar sails weigh around 7 g/m^2 (0.0014 lb/sq ft), giving a thrust of 0.01 mPa (1.5×10^{-9} psi) at 1 AU (150,000,000 km; 93,000,000 mi). This gives a mass/thrust ratio of at least 700 kg/N, similar to a magnetic sail, neglecting other structural components.

The solar and magnetic sails have a thrust that falls off as the square of the distance from the Sun.

When close to a planet with a strong magnetosphere such as Earth or a gas giant, the magnetic sail could generate more thrust by interacting with the magnetosphere instead of the solar wind, and may therefore be more efficient.

Mini-Magnetospheric Plasma Propulsion (M2P2)

In order to reduce the size and weight of the magnet of the magnetic sail, it may be possible to *inflate* the magnetic field using a plasma in the same way that the plasma around the Earth stretches out the Earth's magnetic field in the magnetosphere. In this approach, called mini-magnetospheric plasma propulsion (M2P2), currents that run through the plasma will augment and partially replace the currents in the coil. This is expected to be especially useful far from the Sun, where the increased effective size of a M2P2 sail compensates for the reduced dynamic pressure of the solar wind. The original NASA design proposes a spacecraft containing a can-shaped electromagnet into which a plasma is injected. The plasma pressure stretches the magnetic field and inflates a bubble of plasma around the spacecraft. The plasma then generates a kind of miniaturized magnetosphere around the spacecraft, analogous to the magnetosphere that surrounds the Earth.

The protons and electrons which make up the solar wind are deflected by this magnetosphere and the reaction accelerates the spacecraft. The thrust of the M2P2 device would be steerable to some extent, potentially allowing the spacecraft to 'tack' into the solar wind and allowing efficient changes of orbit.

In the case of the (M2P2) system the spacecraft releases gas to create the plasma needed to maintain the somewhat leaky plasma bubble. The M2P2 system therefore has an effective *specific impulse* which is the amount of gas consumed per newton second of thrust. This is a figure of merit usually used for rockets, where the fuel is actually reaction mass. Robert Winglee, who originally proposed the M2P2 technique, calculates a *specific impulse* of 200 kN·s/kg (roughly 50 times better than the space shuttle main engine).

These calculations suggest that the system requires on the order of a kilowatt of power per newton of thrust, considerably lower than electric thrusters, and that the system generates the same thrust anywhere within the heliopause because the sail spreads automatically as the solar wind becomes less dense. However, this technique is less understood than the simpler magnetic sail and issues of how large and heavy the magnetic coil would have to be or whether the momentum from the solar wind can be efficiently transferred to the spacecraft are under dispute.

The expansion of the magnetic field using plasma injected has been successfully tested in a large vacuum chamber on Earth, but the development of thrust was not part of the experiment. A beam-powered variant, MagBeam, is also under development.

Modes of Operation

In a Plasma Wind

When operating away from planetary magnetospheres, a magnetic sail would force the positively charged protons of the solar wind to curve as they passed through the magnetic field. The change of momentum of the protons would thrust against the magnetic field, and thus against the field coil.

Just as with solar sails, magnetic sails can "tack." If a magnetic sail orients at an angle relative to the solar wind, charged particles are deflected preferentially to one side and the magnetic sail is pushed laterally. This means that magnetic sails could maneuver to most orbits.

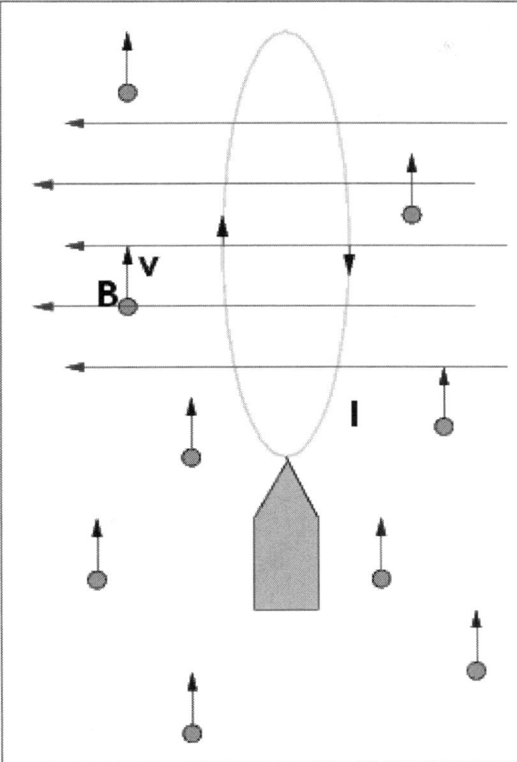

Figure: *A magnetic sail in a wind of charged particles. The sail generates a magnetic field, represented by red arrows, which deflects the particles into the page. The force on the sail is out of the page.*

In this mode, the amount of thrust generated by a magnetic sail falls off with the square of its distance from the Sun as the flux density of charged particles reduces. Solar weather also has major effects on the sail. It is possible that the plasma eruption from a severe solar flare could damage an efficient, fragile sail.

A common misconception is that a magnetic sail cannot exceed the speed of the plasma pushing it. As the speed of a magnetic sail increases, its acceleration becomes more dependent on its ability to tack efficiently. At high speeds, the plasma wind's direction will seem to come increasingly from the front of the spacecraft. Advanced sailing spacecraft might deploy field coils as "keels," so the spacecraft could use the difference in vector between the solar magnetic field and the solar wind, much as sailing yachts do.

Inside a Planetary Magnetosphere

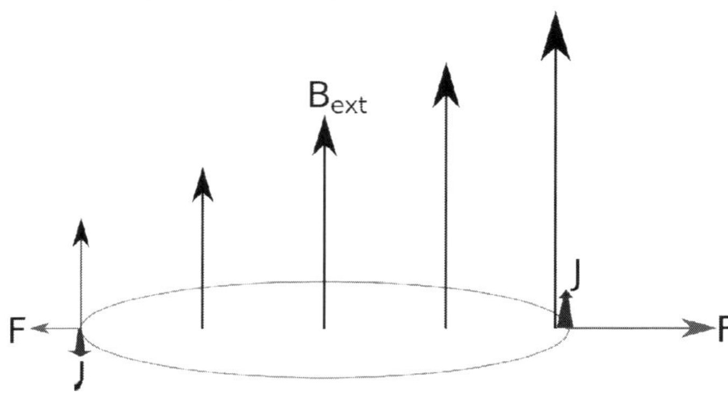

Figure: *A magnetic sail in a spatially varying magnetic field. Because the vertical external field B_{ext} is stronger on one side than the other, the leftward force on the left side of the ring is smaller than the rightward force on the right side of the ring, and the net force on the sail is to the right.*

Inside a planetary magnetosphere, a magnetic sail can thrust against a planet's magnetic field, especially in an orbit that passes over the planet's magnetic poles, in a similar manner to an electrodynamic tether.

The range of maneuvers available to a magnetic sail inside a planetary magnetosphere are more limited than in a plasma wind. Just as with the more familiar small-scale magnets used on Earth, a magnetic sail can only be attracted towards the magnetosphere's poles or repelled from them, depending on its orientation.

When the magnetic sail's field is oriented in the opposite direction to the magnetosphere it experiences a force inward and toward the nearest pole, and when it is oriented in the same direction as the magnetosphere it experiences the opposite effect. A magnetic sail oriented in the same direction as the magnetosphere is not stable, and will have to prevent itself from being flipped over to the opposite orientation by some other means.

The thrust that a magnetic sail delivers within a magnetosphere decreases with the fourth power of its distance from the planet's internal magnetic dynamo.

This limited maneuvering capability is still quite useful. By varying the magnetic sail's field strength over the course of its orbit, a magnetic sail can give itself a "perigee kick" raising the altitude of its orbit's apogee.

Repeating this process with each orbit can drive the magnetic sail's apogee higher and higher, until the magnetic sail is able to leave the planetary magnetosphere and catch the solar wind. The same process in reverse can be used to lower or circularize the apogee of a magsail's orbit when it arrives at a destination planet.

In theory, it is possible for a magnetic sail to launch directly from the surface of a planet near one of its magnetic poles, repelling itself from the planet's magnetic field. However, this requires the magnetic sail to be maintained in its "unstable" orientation. A launch from Earth requires superconductors with 80 times the current density of the best known high-temperature superconductors.

Interstellar Travel

Interstellar space contains very small amounts of hydrogen. A fast-moving sail would ionize this hydrogen by accelerating the electrons in one direction and the oppositely charged protons in the other direction. The energy for the ionization and cyclotron radiation would come from the spacecraft's kinetic energy, slowing the spacecraft. The cyclotron radiation from the acceleration of particles would be an easily detected howl in radio frequencies.

Thus, in interstellar spaceflight outside the heliopause of a star a magnetic sail could act as a parachute to decelerate a spacecraft. This removes any fuel requirements for the deceleration half of an interstellar journey, which would benefit interstellar travel enormously. The magsail was first proposed for this purpose in 1988 by Robert Zubrin and Dana Andrews, predating other uses, and evolved from a concept of the Bussard ramjet which used a magnetic scoop to collect interstellar material.

Magnetic sails could also be used with beam-powered propulsion by using a high-power particle accelerator to fire a beam of charged particles at the spacecraft. The magsail would deflect this beam, transferring momentum to the vehicle. This would provide much higher acceleration than a solar sail driven by a laser, but a charged particle beam would disperse in a shorter distance than a laser due to the electrostatic repulsion of its component particles. This dispersion problem could potentially be resolved by accelerating a stream of sails which then in turn transfer their momentum to a magsail vehicle, as proposed by Jordin Kare.

Fictional Uses

The ancestor of the magsail, the Bussard magnetic scoop, first appeared in science-fiction in Poul Anderson's 1967 short story *To*

Outlive Eternity, which was followed by the novel *Tau Zero* in 1970. The magsail appears as a crucial plot device in *The Children's Hour*, a *Man-Kzin Wars* novel by Jerry Pournelle and S.M. Stirling (1991). It also features prominently in the science-fiction novels of Michael Flynn, particularly in *The Wreck of the River of Stars* (2003); this book is the tale of the last flight of a magnetic sail ship when fusion rockets based on the Farnsworth-Hirsch Fusor have become the preferred technology.

High-Altitude Cusps

Altitude or height is defined based on the context in which it is used (aviation, geometry, geographical survey, sport, and more). As a general definition, altitude is a distance measurement, usually in the vertical or "up" direction, between a reference datum and a point or object. The reference datum also often varies according to the context. Although the term altitude is commonly used to mean the height above sea level of a location, in geography the term elevation is often preferred for this usage.

Vertical distance measurements in the "down" direction are commonly referred to as depth.

Altitude in Aviation

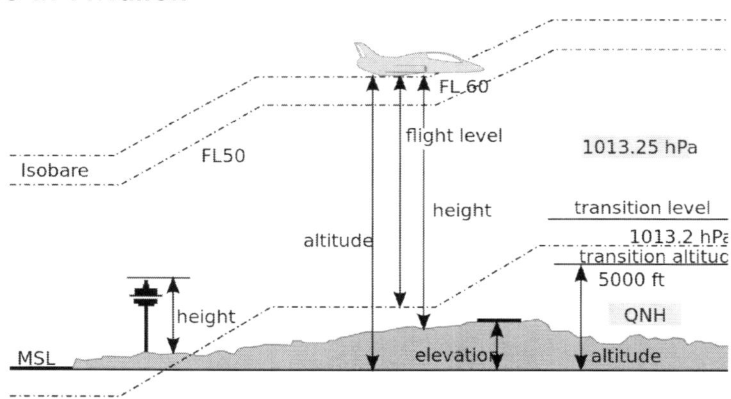

Figure: *Vertical Distance Comparison*

In aviation, the term altitude can have several meanings, and is always qualified by either explicitly adding a modifier (e.g. "true altitude"), or implicitly through the context of the communication. Parties exchanging altitude information must be clear which definition is being used.

Aviation altitude is measured using either Mean Sea Level (MSL) or local ground level (Above Ground Level, or AGL) as the reference datum.

Pressure altitude divided by 100 feet (30 m) as the flight level, and is used above the transition altitude (18,000 feet (5,500 m) in the US, but may be as low as 3,000 feet (910 m) in other jurisdictions); so when the altimeter reads 18,000 ft on the standard pressure setting the aircraft is said to be at "Flight level 180". When flying at a Flight Level, the altimeter is always set to standard pressure (29.92 inHg or 1013.25 hPa).

On the flight deck, the definitive instrument for measuring altitude is the pressure altimeter, which is an aneroid barometer with a front face indicating distance (feet or metres) instead of atmospheric pressure.

There are several types of aviation altitude:

- Indicated altitude is the reading on the altimeter when the altimeter is set to the local barometric pressure at Mean Sea Level.
- Absolute altitude is the height of the aircraft above the terrain over which it is flying. Also referred to feet/metres above ground level (AGL).
- True altitude is the actual elevation above mean sea level. It is Indicated Altitude corrected for non-standard temperature and pressure. In UK aviation radiotelephony usage, *the vertical distance of a level, a point or an object considered as a point, measured from mean sea level*; this is referred to over the radio as altitude.
- Height is the elevation above a ground reference point, commonly the terrain elevation. In UK aviation radiotelephony usage, *the vertical distance of a level, a point or an object considered as a point, measured from a specified datum*; this is referred to over the radio as height, where the specified datum is the airfield elevation.
- Pressure altitude is the elevation above a standard datum air-pressure plane (typically, 1013.25 millibars or 29.92– Hg). Pressure altitude and indicated altitude are the same when the altimeter is set to 29.92" Hg or 1013.25 millibars.
- Density altitude is the altitude corrected for non-ISA International Standard Atmosphere atmospheric conditions. Aircraft performance depends on density altitude, which is affected by barometric pressure, humidity and temperature. On a very hot day, density altitude at an airport (especially one at a high elevation) may be so high as to preclude takeoff, particularly for helicopters or a heavily loaded aircraft.

These types of Altitude can be explained more simply as various ways of measuring the altitude:

- Indicated Altitude – the altimeter reading
- Absolute Altitude – altitude in terms of the distance above the ground directly below
- True altitude – altitude in terms of elevation above sea level
- Height – altitude in terms of the distance above a certain point
- Pressure altitude – the air pressure in terms of altitude in the International Standard Atmosphere
- Density altitude – the density of the air in terms of altitude in the International Standard Atmosphere

Altitude Regions

The Earth's atmosphere is divided into several altitude regions. These regions start and finish at varying heights depending on season and distance from the poles. The altitudes stated below are averages:

- Troposphere — surface to 8,000 metres (5.0 mi) at the poles – 18,000 metres (11 mi) at the equator, ending at the Tropopause.
- Stratosphere — Troposphere to 50 kilometres (31 mi)
- Mesosphere — Stratosphere to 85 kilometres (53 mi)
- Thermosphere — Mesosphere to 675 kilometres (419 mi)
- Exosphere — Thermosphere to 10,000 kilometres (6,200 mi)

High Altitude and Low Pressure

Regions on the Earth's surface (or in its atmosphere) that are high above mean sea level are referred to as high altitude. High altitude is sometimes defined to begin at 2,400 metres (8,000 ft) above sea level.

At high altitude, atmospheric pressure is lower than that at sea level. This is due to two competing physical effects: gravity, which causes the air to be as close as possible to the ground; and the heat content of the air, which causes the molecules to bounce off each other and expand.

Because of the lower pressure, the air expands as it rises, which causes it to cool. Thus, high altitude air is cold, which causes a characteristic alpine climate. This climate dramatically affects the ecology at high altitude.

Relation between temperature and altitude in Earth's atmosphere

The environmental lapse rate (ELR), is the rate of decrease of temperature with altitude in the stationary atmosphere at a given time and location. As an average, the International Civil Aviation Organisation (ICAO) defines an international standard atmosphere (ISA) with a temperature lapse rate of 6.49 K(°C)/1,000 m (3.56 °F or 1.98 K(°C)/1,000 Ft) from sea level to 11 kilometres (36,000 ft). From 11 to 20 kilometres (36,000 to 66,000 ft), the constant temperature is –56.5 °C (–69.7 °F), which is the lowest assumed temperature in the ISA. The standard atmosphere contains no moisture. Unlike the idealized ISA, the temperature of the actual atmosphere does not always fall at a uniform rate with height. For example, there can be an inversion layer in which the temperature increases with height.

Aurora

An aurora is a natural light display in the sky, predominantly seen in the high latitude (Arctic and Antarctic) regions. The name "auroras" is now more commonly used for the linguistic plural "aurorae" of "aurora", so is adopted throughout the main text of this article. Modern style guides recommend that the names of meteorological phenomena, such as *aurora borealis*, be uncapitalized. Auroras are caused by charged particles, mainly electrons and protons, entering the atmosphere from above causing ionisation and excitation of atmospheric constituents, and consequent optical emissions. Incident protons also produce emissions, and convert to hydrogen atoms by gaining an electron from the atmosphere.

Occurrence of Terrestrial Auroras

Most auroras occur in a band known as the *auroral zone* which is typically 3° to 6° wide in latitude and between 10° and 20° from the geomagnetic poles at all local times (or longitudes), but most clearly seen at night against a dark sky. A region displaying an aurora at any given time is known as the auroral oval, a band which is displaced towards the nightside of the Earth. The day-to-day positions of the auroral ovals are posted on the internet. A geomagnetic storm causes the auroral ovals (north and south) to expand, and bring the aurora to lower latitudes.. Early evidence for a geomagnetic connection comes from the statistics of auroral observations. Elias Loomis (1860) and later in more detail Hermann Fritz (1881) and S. Tromholt (1882) established that the aurora appeared mainly in the "auroral zone", an ring-shaped region with a radius of approximately 2500 km around the Earth's magnetic pole. It was hardly ever seen near the geographic pole, which is about 2000 km away from the magnetic pole. The instantaneous distribution of auroras ("auroral oval" is slightly different, centered about 3–5 degrees nightward of the magnetic pole, so that auroral arcs reach furthest toward the equator when the magnetic pole in question is in between the observer and the Sun. The aurora can be seen best at this time, called magnetic midnight.

In northern latitudes, the effect is known as the *aurora borealis* (or the northern lights), named after the Roman goddess of dawn, Aurora, and the Greek name for the north wind, Boreas, by Galileo in 1619. Auroras seen within the auroral oval may be directly overhead, but from farther away they illuminate the poleward horizon as a greenish glow, or sometimes a faint red, as if the Sun were rising from an unusual direction.

Its southern counterpart, the *aurora australis* (or the southern lights), has features that are almost identical to the aurora borealis and changes simultaneously with changes in the northern auroral zone. It is visible from high southern latitudes in Antarctica, South America, New Zealand, and Australia. Auroras also occur on other planets. Similar to the Earth's aurora, they are also visible close to the planet's magnetic poles. Auroras also occur poleward of the auroral zone as either diffuse patches or arcs, which can be sub-visual.

Auroras are occasionally seen in latitudes below the auroral zone, when a geomagnetic storm temporarily enlarges the auroral oval. Large geomagnetic storms are most common during the peak of the eleven-year sunspot cycle or during the three years after the peak. An aurora may appear overhead as a "corona" of rays, diverging as with a meteor shower, from an apparent central location, resulting from perspective. An electron spirals (gyrates) about a field line at an angle that is determined by its velocity vectors, respectively parallel and perpendicular to the local geomagnetic field vector B. This angle is known as the "pitch angle" of the particle.

The distance, or radius, of the electron from the field line at any time is known as its Larmor radius. The pitch angle increases as the electron travels to a region of greater field strength nearer to the atmosphere. Thus it is possible for some particles to return, or mirror, if the angle becomes 90 degrees before entering the atmosphere to collide with the denser molecules there. Other particles, that do not mirror will enter the atmosphere and contribute to the auroral display over a range of altitudes. Other types of auroras have been observed from space, e.g. "poleward arcs" stretching sunward across the polar cap, the related "theta aurora", and "dayside arcs" near noon. These are relatively infrequent and poorly understood. There are other interesting effects such as flickering aurora, "black aurora" and sub-visual red arcs. In addition to all these, a weak glow (often deep red) observed around the two polar cusps, the field lines separating the ones that are closed from those that are swept into the tail.

Images

The altitudes at which auroral emissions occur were revealed by Carl Størmer and his colleagues who used cameras to triangulate more than 12,000 auroras. They discovered that most of the light is produced between 90 and 150 km above the ground, while extending at times to more than 1000 km. Images of auroras are significantly more common today due to the rise of use of digital cameras that have high enough sensitivities. Film and digital exposure to auroral displays is fraught with difficulties, particularly if faithfulness of reproduction is an objective. Due to the different colour spectrum present, and the temporal changes occurring during the exposure, the results are somewhat unpredictable. Different layers of the film emulsion respond differently to lower light levels, and choice of film can be very important. Longer exposures aggregate the rapidly changing energy, and often blanket the dynamic attribute of a display. Higher sensitivity creates issues with graininess.

The aurora frequently appears either as a diffuse glow or as "curtains" that approximately extend in the east-west direction. At some times, they form "quiet arcs"; at others ("active aurora"), they evolve and change constantly. Each curtain consists of many parallel rays, each lined up with the local direction of the magnetic field lines, consistent with auroras being shaped by Earth's magnetic field. In-situ particle measurements confirm that auroral electrons are guided by geomagnetic field lines, and spiral around them while moving toward Earth. The similarity of an auroral display to curtains is often enhanced by folds within the arcs.

David Malin pioneered multiple exposure using multiple filters for astronomical photography, recombining the images in the laboratory to recreate the visual display more accurately. For scientific research, proxies are often used, such as ultra-violet, and re-coloured to simulate the appearance to humans. Predictive techniques are also used, to indicate the extent of the display, a highly useful tool for aurora hunters. Terrestrial features often find their way into aurora images, making them more accessible and more likely to be published by major websites. It is possible to take excellent images with standard film (using ISO ratings between 100 and 400) and a single-lens reflex camera with full aperture, a fast lens (f1.4 50 mm, for example), and exposures between 10 and 30 seconds, depending on the aurora's display strength.

Early work on the imaging of the auroras was done in 1949 by the University of Saskatchewan using the SCR-270 radar.

Visual Forms & Colours

Auroras take many different visual forms. The most distinctive and brightest are the curtain-like auroral arcs. They eventually fragment or 'break-up' into separate, and rapidly changing, often rayed features which may fill the whole sky. These are the 'discrete' auroras which are at times bright enough to read a newspaper by at night. A 'diffuse' aurora, on the other hand, is a relatively featureless glow sometimes close to the limit of visibility. It can be distinguished from moonlit clouds by the fact that stars can be seen undiminished through the glow. Diffuse auroras are composed of patches whose brightness exhibits regular or near-regular pulsations. The pulsation period can be typically many seconds, so is not always obvious. Occasionally there is a fast, sub-second, flickering. A typical auroral display consists of these forms appearing in the above order throughout the night.

- Red: At the highest altitudes, excited atomic oxygen emits at 630.0 nm (red); low concentration of atoms and lower sensitivity of eyes at this wavelength make this colour visible only under some circumstances with more intense solar activity. The low amount of oxygen atoms and their very gradually diminishing concentration is responsible for the faint, gradual appearance of the top parts of the "curtains".
- Green: At lower altitudes the more frequent collisions suppress this mode and the 557.7 nm emission (green) dominates; fairly high concentration of atomic oxygen and higher eye sensitivity in green make green auroras the most common. The excited molecular nitrogen (atomic nitrogen being rare due to high stability of the N_2 molecule) plays its role here as well, as it can transfer energy by collision to an oxygen atom, which then radiates it away at the green wavelength. (Red and green can also mix together to pink or yellow hues.) The rapid decrease of concentration of atomic oxygen below about 100 km is responsible for the abrupt-looking end of the bottom parts of the curtains.
- Yellow and pink are a mix of red and green or blue.
- Blue: At yet lower altitudes atomic oxygen is not common anymore, and ionized molecular nitrogen takes over in visible light emission; it radiates at a large number of wavelengths in both red and blue parts of the spectrum, with 428 nm (blue)

being dominant. Blue and purple emissions, typically at the bottoms of the "curtains", show up at the highest levels of solar activity.

Other Auroral Radiation

In addition, the aurora and associated currents produce a strong radio emission around 150 kHz known as auroral kilometric radiation AKR, first discovered in 1972. Ionospheric absorption makes AKR only observable from space. X-ray emissions, originating from the particles associated with auroras, have also been detected .

Causes of Auroras

A full understanding of the detailed physical processes which lead to different types of auroras is still incomplete, but the basic cause involves the interaction of the solar wind with the earth's magnetosphere. The varying intensity of this produces effects of different magnitudes, but includes one or more of the following physical scenarios.

1. A quiescent solar wind flowing past the Earth's magnetosphere steadily interacts with it and can both inject solar wind particles directly onto the geomagnetic field lines that are 'open', as opposed to 'closed' in the opposite hemisphere, and provide diffusion through the bow shock. It can also cause particles already trapped in the radiation belts to weakly precipitate into the atmosphere. Once particles are lost to the atmosphere from the radiation belts, under quiet conditions new ones only replace them slowly, and the loss-cone becomes depleted. In the magnetotail, however, particle trajectories seem to constantly reshuffle, probably when the particles cross the very weak magnetic field near the equator. As a result, the flow of electrons in that region is nearly the same in all directions ("isotropic"), so assures a steady supply of leaking electrons. The leakage of negative electrons does not leave the tail positively charged, because each leaked electron lost to the atmosphere is quickly replaced by a low energy electron drawn upward from the ionosphere. Such replacement of "hot" electrons by "cold" ones is in complete accord with the 2nd law of thermodynamics. The complete process, which also generates a electrical ring current around the earth, is uncertain.
2. Geomagnetic disturbance from an enhanced solar wind causes distortions of the magnetotail ("magnetic substorms"). These 'substorms' tend to occur after prolonged spells (hours) during

which the interplanetary magnetic field has an appreciable southward component. This leads to a higher rate of interconnection between its field lines and those of Earth. As a result the solar wind moves magnetic flux (tubes of magnetic field lines, 'locked' together with their resident plasma) from the day side of Earth to the magnetotail, widening the obstacle it presents to the solar wind flow and constricting the tail on the night-side. Ultimately some tail plasma can separate ("magnetic reconnection"); some blobs ("plasmoids") are squeezed downstream and are carried away with the solar wind; others are squeezed toward Earth where their motion feeds large outbursts of auroras, mainly around midnight ("unloading process"). A geomagnetic storm resulting from greater interaction adds many more particles to the plasma trapped around Earth, also producing enhancement of the "ring current". Occasionally the resulting modification of the Earth's magnetic field can be so strong that it produces auroras visible at middle latitudes, on field lines much closer to the equator than the auroral zone.

3. Acceleration of auroral charged particles invariably accompanies a magnetospheric disturbance that causes an aurora. This mechanism, which is believed to predominantly arise from wave-particle interactions, raises the velocity of a particle in the direction of the guiding magnetic field. The pitch angle is thereby decreased, and increases the chance of it being precipitated into the atmosphere. Both electromagnetic and electrostatic waves, produced at the time of greater geomagnetic disturbances, make a significant contribution to the energising processes that enable an aurora to be sustained. Particle acceleration provides a complex intermediate process for transferring energy from the solar wind indirectly into the atmosphere.

The details of these phenomena are not fully understood. However it is clear that the prime source of auroral particles is the solar wind feeding the magnetosphere, the reservoir containing radiation belt, and temporarily magnetically trapped, particles confined by the geomagnetic field, coupled with particle acceleration processes.

Auroral Particles

The local cause of the ionization and excitation of atmospheric constituents leading to auroral emissions was discovered in 1960, with a pioneering rocket flight made from Fort Churchill in Canada, to be a flux of electrons entering the atmosphere from above. Since then an extensive collection of measurements has been acquired painstakingly and with steadily improving resolution since the 1960s by many research teams using rockets and satellites to traverse the auroral zone. The main findings have been that auroral arcs and other bright forms are due to electrons which have been accelerated during the final few 10,000 km or so of their plunge into the atmosphere. These electrons often, but not always, exhibit a peak in their energy distribution, and are preferentially aligned along the local direction of the magnetic field. The electrons which are mainly responsible for diffuse and pulsating auroras have, in contrast, a smoothly falling energy distribution, and an angular (pitch-angle) distribution favouring directions perpendicular to the local magnetic field. Pulsations were discovered to originate at or close to the equatorial crossing point of auroral zone magnetic field lines. Protons are also associated with auroras, particularly where the radiation belts predominate towards lower latitudes.

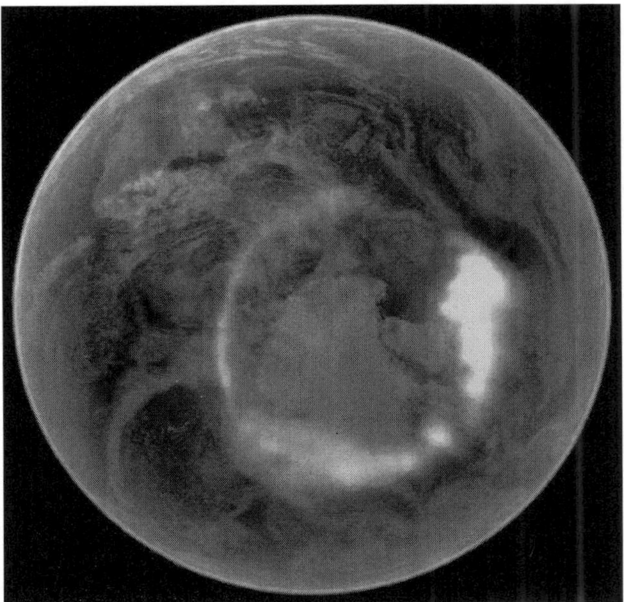

Figure: *Aurora australis (11 September 2005) as captured by NASA's IMAGE satellite, digitally overlaid onto The Blue Marble composite image. An animation created using the same satellite data is also available*

Auroras and the Atmosphere

Auroras result from emissions of photons in the Earth's upper atmosphere, above 80 km (50 mi), from ionized nitrogen atoms regaining an electron, and oxygen atoms and nitrogen based molecules returning from an excited state to ground state. They are ionized or excited by the collision of particles precipitated into the atmosphere. Both incoming electrons and protons may be involved. Excitation energy is lost within the atmosphere by the emission of a photon, or by collision with another atom or molecule: oxygen emissions green or orange-red, depending on the amount of energy absorbed. Nitrogen emissions blue or red; blue if the atom regains an electron after it has been ionized, red if returning to ground state from an excited state.

Oxygen is unusual in terms of its return to ground state: it can take three quarters of a second to emit green light and up to two minutes to emit red. Collisions with other atoms or molecules absorb the excitation energy and prevent emission. Because the very top of the atmosphere has a higher percentage of oxygen and is sparsely distributed such collisions are rare enough to allow time for oxygen to emit red. Collisions become more frequent progressing down into the atmosphere, so that red emissions do not have time to happen, and eventually even green light emissions are prevented. This is why there is a colour differential with altitude; at high altitude oxygen red dominates, then oxygen green and nitrogen blue/red, then finally nitrogen blue/red when collisions prevent oxygen from emitting anything. Green is the most common of all auroras. Behind it is pink, a mixture of light green and red, followed by pure red, yellow (a mixture of red and green), and finally, pure blue.

Auroras and the Ionosphere

Bright auroras are generally associated with Birkeland currents (Schield et al., 1969; Zmuda and Armstrong, 1973) which flow down into the ionosphere on one side of the pole and out on the other. In between, some of the current connects directly through the ionospheric E layer (125 km); the rest ("region 2") detours, leaving again through field lines closer to the equator and closing through the "partial ring current" carried by magnetically trapped plasma. The ionosphere is an ohmic conductor, so such currents require a driving voltage, which some dynamo mechanism can supply. Electric field probes in orbit above the polar cap suggest voltages of the order of 40,000 volts, rising up to more than 200,000 volts during intense magnetic storms. Ionospheric resistance has a complex nature, and leads to a secondary Hall current flow. By a strange twist of physics, the magnetic disturbance on the ground due to the main current almost cancels out, so most of the observed effect of auroras is due to a secondary current, the auroral electrojet. An auroral electrojet index (measured in nanotesla) is regularly derived from ground data and serves as a general measure of auroral activity. Kristian Birkeland deduced that the currents flowed in the east-west directions along the auroral arc, and such currents, flowing from the dayside toward (approximately) midnight were later named "auroral electrojets".

Interaction of the Solar Wind with Earth

The Earth is constantly immersed in the solar wind, a rarefied flow of hot plasma (gas of free electrons and positive ions) emitted by the Sun in all directions, a result of the two-million-degree heat of the Sun's outermost layer, the corona. The solar wind usually reaches Earth with a velocity around 400 km/s, density around 5 ions/cm^3 and magnetic field intensity around 2–5 nT (nanoteslas; Earth's surface field is typically 30,000–50,000 nT). These are typical values. During magnetic storms, in particular, flows can be several times faster; the interplanetary magnetic field (IMF) may also be much stronger. Astrophysicist Joan Feynman deduced in the 1970s that the long-term averages of solar wind speed correlated with geomagnetic activity. Her work resulted from data collected by the Explorer 33 spacecraft The solar wind and magnetosphere consist of plasma (ionized gas), which conducts electricity. It is well known (since Michael Faraday's work around 1830) that when an electrical conductor is placed within a magnetic field while relative motion occurs in a direction that the conductor cuts *across* (or is cut *by*), rather than *along*, the lines of the magnetic field, an electric current is said to be induced into that conductor and electrons flow within it.

The amount of current flow is dependent upon a) the rate of relative motion, b) the strength of the magnetic field, c) the number of conductors ganged together and d) the distance between the conductor and the magnetic field, while the *direction* of flow is dependent upon the direction of relative motion. Dynamos make use of this basic process ("the dynamo effect"), any and all conductors, solid or otherwise are so affected, including plasmas or other fluids. The IMF originates on the Sun, linked to the sunspots, and its field lines (lines of force) are dragged out by the solar wind. That alone would tend to line them up in the Sun-Earth direction, but the rotation of the Sun angles them at Earth by about 45 degrees forming a spiral in the ecliptic plane), known as the spiral shape after the physicist.

The field lines passing Earth will therefore usually be linked to those near the western edge ("limb") of the visible Sun at any time. The solar wind and the magnetosphere, being two electrically conducting fluids in relative motion, should be able in principle to generate electric currents by dynamo action and impart energy from the flow of the solar wind. However, this process is hampered by the fact that plasmas conduct easily along magnetic field lines, but not so easily perpendicular to them. Energy is more effectively transferred by temporary magnetic connection between the field lines of the solar wind and those of the magnetosphere. Unsurprisingly this process is known as magnetic reconnection. As already said, it happens most readily when the interplanetary field is directed southward, closer to that of the geomagnetic field in the regions of both the north magnetic pole and south magnetic pole.

Auroras are more frequent and brighter during the intense phase of the solar cycle when coronal mass ejections increase the intensity of the solar wind.

Magnetosphere

Earth's magnetosphere is shaped by the impact of the solar wind on the Earth's magnetic field which forms an obstacle to the flow, diverting it, at an average distance of about 70,000 km (11 Earth radii or Re), producing a bow shock 12,000 km to 15,000 km (1.9 to 2.4 Re) further upstream. The width of the magnetosphere abreast of Earth, is typically 190,000 km (30 Re), and on the night side a long "magnetotail" of stretched field lines extends to great distances (> 200 Re). The high latitude magnetosphere is filled with plasma as the solar wind passes the Earth. The flow of plasma into the magnetosphere increases with solar wind density and speed , favoured by a southward component of the IMF and additional turbulence in the solar wind flow. The flow pattern of magnetospheric plasma is mainly from the magnetotail toward the Earth, around the Earth and back into the solar wind through the magnetopause on the day-side. In addition to moving perpendicular to the Earth's magnetic field, some magnetospheric plasma travels down along the Earth's magnetic field lines, gains additional energy and loses it to the atmosphere in the auroral zones.

The cusps of the magnetosphere, separating open from closed geomagnetic field lines allow a small amount of solar wind to directly reach the top of the atmosphere, producing an auroral glow. The electrons responsible for the brightest forms of aurora are well accounted for by their acceleration in the dynamic electric fields of plasma turbulence encountered during precipitation from the magnetosphere into the auroral atmosphere. In contrast, static electric fields are unable to transfer energy to the electrons due to their conservative nature.

The un-accelerated electrons and ions cause the dim glow of the diffuse aurora. On 26 February 2008, THEMIS probes were able to determine, for the first time, the triggering event for the onset of magnetospheric substorms. Two of the five probes, positioned approximately one third the distance to the moon, measured events suggesting a magnetic reconnection event 96 seconds prior to auroral intensification.

Geomagnetic Storms

Geomagnetic storms that ignite auroras actually happen more often during the months around the equinoxes. It is not well understood why geomagnetic storms are tied to Earth's seasons, but it is known that during spring and autumn, the interplanetary magnetic field and that of Earth link up. At the magnetopause, Earth's magnetic field points north. When B_z becomes large and negative (i.e., the IMF tilts south), it can partially cancel Earth's magnetic field at the point of contact. South-pointing B_zs open a door through which energy from the solar wind reaches Earth's inner magnetosphere. The peaking of B_z during this time is a result of geometry. The IMF from the Sun is carried outward with the solar wind and the southward (and northward) excursions of B_z are greatest during April and October, when Earth's magnetic dipole axis is most closely aligned with the Parker spiral.

B_z is not the only influence on geomagnetic activity, however, the Sun's rotation axis is tilted 8 degrees with respect to the plane of Earth's orbit. The solar wind blows more rapidly from the Sun's poles than from its equator, thus the average speed of particles buffeting Earth's magnetosphere waxes and wanes every six months. The solar wind speed is greatest – by about 50 km/s, on average – around 5 September and 5 March when Earth lies at its highest heliographic latitude.

Still, neither B_z nor the solar wind can fully explain the seasonal behaviour of geomagnetic storms. Those factors together contribute only about one-third of the observed semi-annual variations.

Auroral Particle Acceleration

The electrons responsible for the brightest forms of aurora are well accounted for by their acceleration in the dynamic electric fields of plasma turbulence encountered during precipitation from the magnetosphere into the auroral atmosphere. In contrast, static electric fields are unable to transfer energy to the electrons due to their conservative nature. The un-accelerated electrons and ions cause the dim glow of the diffuse aurora.

The convergence of magnetic field lines towards the Earth that creates a 'magnetic mirror' turns back many of the downward flowing electrons as the field strength increases. The bright forms of auroras are produced when downward acceleration not only increases the energy of precipitating electrons but also reduces their pitch angles (angle between electron velocity and the local magnetic field vector). This greatly increases the rate of deposition of energy into the atmosphere, and thereby the rates of ionisation, excitation and consequent auroral light emission. One early theory proposed for the acceleration of auroral electrons is based on an assumed static, or quasi-static, electric field and a consequent uni-directional potential drop.

The originating charge assembly and associated equi-potentials are so-far unspecified. Poisson's equation indicates that there can be no configuration of charge resulting in a net potential drop. This fact prohibits the concept of a sustained uni-directional potential drop along the Earth's magnetic field lines. The electric field theory proposed for auroral particle acceleration is therefore highly questionable as it appears to violate a basic principle of physics. A more credible theory is based on acceleration by Landau resonance in the turbulent electric fields of the acceleration region. This process is essentially the same as that employed in plasma fusion laboratories throughout the world, and appears well able to account in principle for most – if not all – detailed properties of the electrons responsible for the brightest forms of aurorae.

Other mechanisms have also been proposed, in particular, Alfvén waves, wave modes involving the magnetic field first noted by Hannes Alfvén (1942), which have been observed in the lab and in space. The question is whether these waves might just be a different way of looking at the above process, however, because this approach does not point out a different energy source, and many plasma bulk phenomena can also be described in terms of Alfvén waves.

Other processes are also involved in the aurora, and much remains to be learned. Auroral electrons created by large geomagnetic storms often seem to have energies below 1 keV, and are stopped higher up, near 200 km. Such low energies excite mainly the red line of oxygen, so that often such auroras are red. On the other hand, positive ions also reach the ionosphere at such time, with energies of 20–30 keV, suggesting they might be an "overflow" along magnetic field lines of the copious "ring current" ions accelerated at such times, by processes different from the ones described above. Some O+ ions ("conics") also seem accelerated in different ways by plasma processes associated with the aurora. These ions are accelerated by plasma waves, in directions mainly perpendicular to the field lines. They therefore start at their own "mirror points" and can travel only upward. As they do so, the "mirror effect" transforms their directions of motion, from perpendicular to the line to lying on a cone around it, which gradually narrows down.

Auroral Events of Historical Significance

The auroras that resulted from the "great geomagnetic storm" on both 28 August and 2 September 1859 are thought the most spectacular in recent recorded history. In a paper to the Royal Society on 21 November 1861, Scottish physicist Balfour Stewart described both auroral events as documented by a self-recording magnetograph at the Kew Observatory and established the connection between the 2 September 1859 auroral storm and the Carrington-Hodgson flare event when he observed that, "It is not impossible to suppose that in this case our luminary was taken *in the act*."

The second auroral event, which occurred on 2 September 1859 as a result of the exceptionally intense Carrington-Hodgson white light solar flare on 1 September 1859, produced auroras, so widespread and extraordinarily bright, that they were seen and reported in published scientific measurements, ship logs, and newspapers throughout the United States, Europe, Japan, and Australia. It was reported by the *New York Times* that in Boston on Friday 2 September 1859 the aurora was "so brilliant that at about one o'clock ordinary print could be read by the light". One o'clock EST time on Friday 2 September, would have been 6:00 GMT and the self-recording magnetograph at the Kew Observatory was recording the geomagnetic storm, which was then one hour old, at its full intensity. Between 1859 and 1862, Elias Loomis published a series of nine papers on the Great Auroral Exhibition of 1859 in the *American Journal of Science* where he collected world-wide reports of the auroral event.

That aurora is thought to have been produced by one of the most intense coronal mass ejections in history. It is also notable for the fact that it is the first time where the phenomena of auroral activity and electricity were unambiguously linked. This insight was made possible not only due to scientific magnetometer measurements of the era, but also as a result of a significant portion of the 125,000 miles (201,000 km) of telegraph lines then in service being significantly disrupted for many hours throughout the storm. Some telegraph lines, however, seem to have been of the appropriate length and orientation to produce a sufficient geomagnetically induced current from the electromagnetic field to allow for continued communication with the telegraph operator power supplies switched off.

Historical Theories, Superstition and Mythology

Magnetic control of the aurora was mentioned by Ancient Greek explorer/geographer Pytheas, Hiorter, and Celsius described in 1741 evidence that large magnetic fluctuations occurred whenever the aurora was observed overhead. It was also later realised that large electric currents were associated with the aurora, flowing in the region where auroral light originated. Multiple superstitions and obsolete theories explaining the aurora have emerged over the centuries.

- Seneca speaks diffusely on auroras in the first book of his Naturales Quaestiones, drawing mainly from Aristotle; he classifies them "putei" or wells when they are circular and "rim a large hole in the sky", "pithaei" when they look like casks, "chasmata" from the same root of the English chasm, "pogoniae" when they are bearded, "cyparissae" when they look like cypresses), describes their manifold colours and asks himself whether they are above or below the clouds. He recalls that under Tiberius, an aurora formed above Ostia, so intense and so red that a cohort of the army, stationed nearby for fireman duty, galloped to the city.
- Walter William Bryant wrote in his book *Kepler* (1920) that Tycho Brahe "seems to have been something of a homœopathist, for he recommends sulfur to cure infectious diseases "brought on by the sulphurous vapours of the Aurora Borealis."
- Benjamin Franklin theorized that the "mystery of the Northern Lights" was caused by a concentration of electrical charges in the polar regions intensified by the snow and other moisture.

The northern lights have had a number of names throughout history. The Cree called the phenomenon the "Dance of the Spirits". In Medieval Europe, the auroras were commonly believed to be a sign from God.

There is the Claim from 1855 that in Norse Mythology

The Valkyrior are warlike virgins, mounted upon horses and armed with helmets and spears. /.../ When they ride forth on their errand, their armour sheds a strange flickering light, which flashes up over the northern skies, making what Men call the "aurora borealis", or "Northern Lights".

While a striking notion, there is not a vast body of evidence in the Old Norse literature giving this interpretation, or even much reference to auroras. Although auroral activity is common over Scandinavia and Iceland today, it is possible that the Magnetic North Pole was considerably farther away from this region during the relevant period of Norse mythology, and .

The first Old Norse account of *norπrljσs* is found in the Norwegian chronicle *Konungs Skuggsjα* from AD 1230. The chronicler has heard about this phenomenon from compatriots returning from Greenland, and he gives three possible explanations: that the ocean was surrounded by vast fires, that the sun flares could reach around the world to its night side, or that glaciers could store energy so that they eventually became fluorescent.

In ancient Roman mythology, Aurora is the goddess of the dawn, renewing herself every morning to fly across the sky, announcing the arrival of the sun. The persona of Aurora the goddess has been incorporated in the writings of Shakespeare, Lord Tennyson, and Thoreau.

In the traditions of Aboriginal Australians, the Aurora Australis is commonly associated with fire. For example, the Gunditjmara people of western Victoria called auroras "Puae buae", meaning "ashes", while the Gunai people of eastern Victoria perceived auroras as bushfires in the spirit world. When the Dieri people of South Australia said that an auroral display was "Kootchee", an evil spirit creating a large fire. Similarly, the Ngarrindjeri people of South Australia referred to auroras seen over Kangaroo Island as the campfires of spirits in the 'Land of the Dead'. Aboriginal people in southwest Queensland believed the auroras to be the fires of the "Oola Pikka", ghostly spirits who spoke to the people through auroras. Sacred law forbade anyone except male elders from watching or interpreting the messages of ancestors they believed were transmitted through auroras.

After the Battle of Fredericksburg, the lights could be seen from the battlefield that night. The Confederate Army took it as a sign that God was on their side during the battle as it was very rare that one could see the Lights in Virginia. The painting Aurora Borealis (1865) by American landscape painter Frederic Edwin Church is widely interpreted to represent the conflict of the American Civil War.

Planetary Auroras

Both Jupiter and Saturn have magnetic fields much stronger than Earth's (Jupiter's equatorial field strength is 4.3 gauss, compared to

0.3 gauss for Earth), and both have large radiation belts. Auroras have been observed on both, most clearly with the Hubble Space Telescope. Uranus and Neptune have also been observed to have auroras.

The auroras on the gas giants seem, like Earth's, to be powered by the solar wind. In addition, however, Jupiter's moons, especially Io, are powerful sources of auroras on Jupiter. These arise from electric currents along field lines ("field aligned currents"), generated by a dynamo mechanism due to the relative motion between the rotating planet and the moving moon. Io, which has active volcanism and an ionosphere, is a particularly strong source, and its currents also generate radio emissions, studied since 1955. Auroras also have been observed on the surfaces of Io, Europa, and Ganymede, using the Hubble Space Telescope.

These auroras have also been observed on Venus and Mars. Because Venus has no intrinsic (planetary) magnetic field, Venusian auroras appear as bright and diffuse patches of varying shape and intensity, sometimes distributed across the full planetary disc. Venusian auroras are produced by the impact of electrons originating from the solar wind and precipitating in the night-side atmosphere. An aurora was also detected on Mars, on 14 August 2004, by the SPICAM instrument aboard Mars Express. The aurora was located at Terra Cimmeria, in the region of 177° East, 52° South.

The total size of the emission region was about 30 km across, and possibly about 8 km high. By analyzing a map of crustal magnetic anomalies compiled with data from Mars Global Surveyor, scientists observed that the region of the emissions corresponded to an area where the strongest magnetic field is localized. This correlation indicates that the origin of the light emission was a flux of electrons moving along the crust magnetic lines and exciting the upper atmosphere of Mars.

Effects of High Altitude on Humans

Medicine recognises that altitudes above 1,500 metres (4,900 ft) start to affect humans, and there is no record of humans living at extreme altitudes above 5,500–6,000 metres (18,000–19,700 ft) for more than two years. As the altitude increases, atmospheric pressure decreases, which affects humans by reducing the partial pressure of oxygen. The lack of oxygen above 2,400 metres (8,000 ft) can cause serious illnesses such as altitude sickness, high altitude pulmonary edema, and high altitude cerebral edema. The higher the altitude, the more likely are serious effects. The human body can adapt to high altitude by breathing faster, having a higher heart rate, and adjusting its blood chemistry. It can take days or weeks to adapt to high altitude. However, above 8,000 metres (26,000 ft), (in the "death zone"), altitude acclimatization becomes impossible.

There is a significantly lower overall mortality rate for permanent residents at higher altitudes. Additionally, there is a dose response relationship between increasing elevation and decreasing obesity prevalence in the United States. In addition, the recent hypothesis suggests that high altitude could be protective against Alzheimer's disease via action of erythropoietin, a hormone released by kindney in response to hypoxia. However, people living at higher elevations have a statistically significant higher rate of suicide. The cause for the increased suicide risk is unknown so far.

Athletes

For athletes, high altitude produces two contradictory effects on performance. For explosive events (sprints up to 400 metres, long jump, triple jump) the reduction in atmospheric pressure signifies less atmospheric resistance, which generally results in improved athletic performance. For endurance events (races of 5,000 metres or more) the predominant effect is the reduction in oxygen which generally reduces the athlete's performance at high altitude. Sports organisations acknowledge the effects of altitude on performance: the International Association of Athletic Federations (IAAF), for example, marks record performances achieved at an altitude greater than 1,000 metres (3,300 ft) with the letter "A".

Athletes also can take advantage of altitude acclimatization to increase their performance. The same changes that help the body cope with high altitude increase performance back at sea level. These changes are the basis of altitude training which forms an integral part of the training of athletes in a number of endurance sports including track and field, distance running, triathlon, cycling and swimming.

Effect of High Altitude on Animals

Decreased oxygen availability and decreased temperature make life at high altitude challenging. Despite these environmental conditions, many species have been successfully adapted at high altitudes. Animals have developed physiological adaptations to enhance oxygen uptake and delivery to tissues which can be used to sustain metabolism. The strategies used by animals to adapt to high altitude depend on their morphology and phylogeny.

Fish

Fish at high altitudes may also have a lower metabolic rate, as has been shown in highland westslope cutthroat trout compared to introduced lowland rainbow trout in the Oldman River basin. There is also a general trend of smaller body sizes and lower species richness at high altitudes observed in aquatic invertebrates, likely due to lower oxygen partial pressures. These factors may decrease productivity in high altitude habitats, meaning there will be less energy available for consumption, growth, and activity, which provides an advantage to fish with lower metabolic demands.

The naked carp from Lake Qinghai, like other members of the carp family, can use gill remodelling to increase oxygen uptake in hypoxia. The response of naked carp to cold and low-oxygen conditions seem to be at least partly mediated by hypoxia-inducible factor 1 (HIF-1). It is unclear whether this is a common characteristic in other high altitude dwelling fish or if gill remodelling and HIF-1 use for cold adaptation are limited to carp.

Rodents

Rodents living at high altitude include deer mice, guinea pigs and rats. As small mammals they face the challenge of maintaining body heat in cold temperatures, due to their large volume to surface area ratio. As oxygen is used as a source of metabolic heat production, the hypobaric hypoxia at high altitudes is problematic.

There are a number of mechanisms that help them survive these harsh conditions including altered genetics of the hemoglobin gene in guinea pigs and deer mice. Deer mice use a high percentage of fats as metabolic fuel at high altitude to retain carbohydrates for small burst of energy. To convert fats to energy in the form of ATP, more oxygen is required than to convert the same amount of carbohydrates. The reason they use fats is believed to be because they have it in large stores, but also means that they must eat more or they will begin to lose weight.

Other physiological changes that occur in rodents at high altitude include increased breathing rate and altered morphology of the lungs and heart allowing more efficient gas exchange and delivery. Lungs of high altitude mice are larger, with more capillaries, and hearts of mice and rats at high altitude have a heavier right ventricle, which pumps blood to the lungs.

At high altitudes, some rodents even shift their thermal neutral zone so they may maintain normal basal metabolic rate at colder temperatures.

Birds

Birds have been especially successful at living at high altitudes. In general, birds have physiological features that are advantageous for high-altitude flight. The respiratory system of birds moves oxygen across the pulmonary surface during both inhalation and exhalation, making it more efficient than that of mammals. In addition, the air circulates in one direction through the parabronchioles in the lungs. Parabronchioles are oriented perpendicular to the pulmonary arteries, forming a cross-current gas exchanger. This arrangement allows for more oxygen to be extracted compared to mammalian concurrent gas exchange; as oxygen diffuses down its concentration gradient and the air gradually becomes more deoxygenated, the pulmonary arteries are still able to extract oxygen. Birds also have a high capacity for oxygen delivery to the tissues because they have larger hearts and cardiac stroke volume (mL / min) compared to mammals of similar body size. Additionally, they have an increased vascularization in flight muscle due to increased branching of capillaries and small muscle fibres (which increases surface-area-to-volume ratio). These two features facilitate oxygen diffusion from the blood to muscle, allowing flight to be sustained during environmental hypoxia. Bird's hearts and brains, which are very sensitive to arterial hypoxia, are more vascularized compared to mammals. The bar-headed goose (*Anser indicus*) is an iconic high flyer that surmounts the Himalayas during migration, and serves as a model system for derived physiological adaptations for high-altitude flight.

Chapter 5

Boundary Layer

In the Earth's atmosphere, the atmospheric boundary layer is the air layer near the ground affected by diurnal heat, moisture or momentum transfer to or from the surface. On an aircraft wing the boundary layer is the part of the flow close to the wing, where viscous forces distort the surrounding non-viscous flow. The thin shear layer which develops on an oscillating body is an example of a Stokes boundary layer, while the Blasius boundary layer refers to the well-known similarity solution near an attached flat plate held in an oncoming unidirectional flow. When a fluid rotates and viscous forces are balanced by the Coriolis effect (rather than convective inertia), an Ekman layer forms. In the theory of heat transfer, a thermal boundary layer occurs. A surface can have multiple types of boundary layer simultaneously.

Aerodynamics

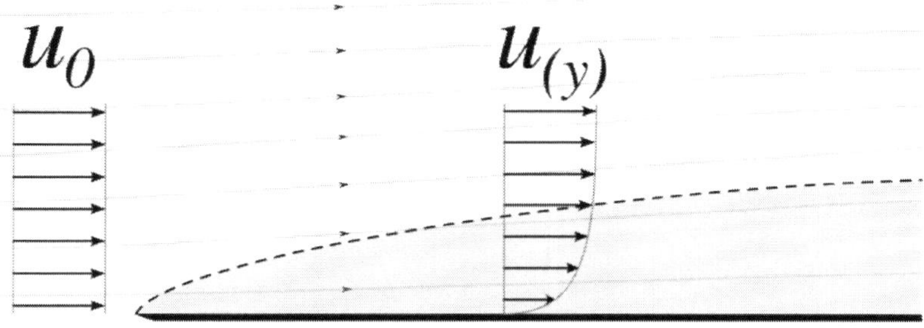

Figure: *Laminar boundary layer velocity profile*

The aerodynamic boundary layer was first defined by Ludwig Prandtl in a paper presented on August 12, 1904 at the third

International Congress of Mathematicians in Heidelberg, Germany. It simplifies the equations of fluid flow by dividing the flow field into two areas: one inside the boundary layer, dominated by viscosity and creating the majority of drag experienced by the boundary body; and one outside the boundary layer, where viscosity can be neglected without significant effects on the solution. This allows a closed-form solution for the flow in both areas, a significant simplification of the full Navier–Stokes equations. The majority of the heat transfer to and from a body also takes place within the boundary layer, again allowing the equations to be simplified in the flow field outside the boundary layer. The pressure distribution throughout the boundary layer in the direction normal to the surface (such as an airfoil) remains constant throughout the boundary layer, and is the same as on the surface itself.

The thickness of the velocity boundary layer is normally defined as the distance from the solid body at which the viscous flow velocity is 99% of the freestream velocity (the surface velocity of an inviscid flow). Displacement Thickness is an alternative definition stating that the boundary layer represents a deficit in mass flow compared to inviscid flow with slip at the wall. It is the distance by which the wall would have to be displaced in the inviscid case to give the same total mass flow as the viscous case. The no-slip condition requires the flow velocity at the surface of a solid object be zero and the fluid temperature be equal to the temperature of the surface. The flow velocity will then increase rapidly within the boundary layer, governed by the boundary layer equations, below.

The thermal boundary layer thickness is similarly the distance from the body at which the temperature is 99% of the temperature found from an inviscid solution. The ratio of the two thicknesses is governed by the Prandtl number. If the Prandtl number is 1, the two boundary layers are the same thickness. If the Prandtl number is greater than 1, the thermal boundary layer is thinner than the velocity boundary layer. If the Prandtl number is less than 1, which is the case for air at standard conditions, the thermal boundary layer is thicker than the velocity boundary layer.

In high-performance designs, such as gliders and commercial aircraft, much attention is paid to controlling the behaviour of the boundary layer to minimize drag. Two effects have to be considered. First, the boundary layer adds to the effective thickness of the body, through the displacement thickness, hence increasing the pressure drag. Secondly, the shear forces at the surface of the wing create skin friction drag.

At high Reynolds numbers, typical of full-sized aircraft, it is desirable to have a laminar boundary layer. This results in a lower skin friction due to the characteristic velocity profile of laminar flow. However, the boundary layer inevitably thickens and becomes less stable as the flow develops along the body, and eventually becomes turbulent, the process known as boundary layer transition. One way of dealing with this problem is to suck the boundary layer away through a porous surface. This can reduce drag, but is usually impractical due to its mechanical complexity and the power required to move the air and dispose of it. Natural laminar flow techniques push the boundary layer transition aft by reshaping the aerofoil or fuselage so that its thickest point is more aft and less thick. This reduces the velocities in the leading part and the same Reynolds number is achieved with a greater length.

At lower Reynolds numbers, such as those seen with model aircraft, it is relatively easy to maintain laminar flow. This gives low skin friction, which is desirable. However, the same velocity profile which gives the laminar boundary layer its low skin friction also causes it to be badly affected by adverse pressure gradients. As the pressure begins to recover over the rear part of the wing chord, a laminar boundary layer will tend to separate from the surface. Such flow separation causes a large increase in the pressure drag, since it greatly increases the effective size of the wing section.

In these cases, it can be advantageous to deliberately trip the boundary layer into turbulence at a point prior to the location of laminar separation, using a turbulator. The fuller velocity profile of the turbulent boundary layer allows it to sustain the adverse pressure gradient without separating.

Thus, although the skin friction is increased, overall drag is decreased. This is the principle behind the dimpling on golf balls, as well as vortex generators on aircraft. Special wing sections have also been designed which tailor the pressure recovery so laminar separation is reduced or even eliminated. This represents an optimum compromise between the pressure drag from flow separation and skin friction from induced turbulence.

When using half-models in wind tunnels, a peniche is sometimes used to reduce or eliminate the effect of the boundary layer.

Naval Architecture

Many of the principles that apply to aircraft also apply to ships, submarines, and offshore platforms.

For ships, unlike aircraft, one deals with incompressible flows, where change in water density is negligible (a pressure rise close to 1000kPa leads to a change of only 2–3 kg/m^3). This field of fluid dynamics is called hydrodynamics. A ship engineer designs for hydrodynamics first, and for strength only later. The boundary layer development, breakdown, and separation become critical because the high viscosity of water produces high shear stresses. Another consequence of high viscosity is the slip stream effect, in which the ship moves like a spear tearing through a sponge at high velocity.

Boundary Layer Equations

The deduction of the boundary layer equations was one of the most important advances in fluid dynamics (Anderson, 2005). Using an order of magnitude analysis, the well-known governing Navier–Stokes equations of viscous fluid flow can be greatly simplified within the boundary layer. Notably, the characteristic of the partial differential equations (PDE) becomes parabolic, rather than the elliptical form of the full Navier–Stokes equations. This greatly simplifies the solution of the equations. By making the boundary layer approximation, the flow is divided into an inviscid portion (which is easy to solve by a number of methods) and the boundary layer, which is governed by an easier to solve PDE. The continuity and Navier–Stokes equations for a two-dimensional steady incompressible flow in Cartesian coordinates are given by

$$\frac{\partial u}{\partial x} + \frac{\partial \upsilon}{\partial y} = 0$$

$$u\frac{\partial u}{\partial x} + \upsilon\frac{\partial u}{\partial y} = -\frac{1}{\rho}\frac{\partial p}{\partial x} + \nu\left(\frac{\partial^2 u}{\partial x^2} + \frac{\partial^2 u}{\partial y^2}\right)$$

$$u\frac{\partial \upsilon}{\partial x} + \upsilon\frac{\partial \upsilon}{\partial y} = -\frac{1}{\rho}\frac{\partial p}{\partial y} + \nu\left(\frac{\partial^2 \upsilon}{\partial x^2} + \frac{\partial^2 \upsilon}{\partial y^2}\right)$$

where u and υ are the velocity components, ρ is the density, p is the pressure, and υ is the kinematic viscosity of the fluid at a point.

The approximation states that, for a sufficiently high Reynolds number the flow over a surface can be divided into an outer region of inviscid flow unaffected by viscosity (the majority of the flow), and a region close to the surface where viscosity is important (the boundary layer). Let u and υ be streamwise and transverse (wall normal) velocities respectively inside the boundary layer. Using scale analysis, it can be shown that the above equations of motion reduce within the boundary layer to become

$$\frac{\partial u}{\partial x}+\frac{\partial \upsilon}{\partial y}=0$$

$$u\frac{\partial u}{\partial x}+\upsilon\frac{\partial u}{\partial y}=-\frac{1}{\rho}\frac{\partial p}{\partial x}+\nu\frac{\partial^2 u}{\partial y^2}$$

and if the fluid is incompressible (as liquids are under standard conditions):

$$\frac{1}{\rho}\frac{\partial p}{\partial y}=0$$

The asymptotic analysis also shows that υ, the wall normal velocity, is small compared with u the streamwise velocity, and that variations in properties in the streamwise direction are generally much lower than those in the wall normal direction.

Since the static pressure p is independent of y, then pressure at the edge of the boundary layer is the pressure throughout the boundary layer at a given streamwise position. The external pressure may be obtained through an application of Bernoulli's equation. Let u_0 be the fluid velocity outside the boundary layer, where and are both parallel. This gives upon substituting for p the following result

$$u\frac{\partial u}{\partial x}+\upsilon\frac{\partial u}{\partial y}=u_0\frac{\partial u_0}{\partial x}+\nu\frac{\partial^2 u}{\partial y^2}$$

with the boundary condition

$$\frac{\partial u}{\partial x}+\frac{\partial v}{\partial y}=0$$

For a flow in which the static pressure p also does not change in the direction of the flow then

$$\frac{\partial p}{\partial x}=0$$

so u_0 remains constant.

Therefore, the equation of motion simplifies to become

$$u\frac{\partial u}{\partial x}+\upsilon\frac{\partial u}{\partial y}=\nu\frac{\partial^2 u}{\partial y^2}$$

These approximations are used in a variety of practical flow problems of scientific and engineering interest. The above analysis is for any instantaneous laminar or turbulent boundary layer, but is used mainly in laminar flow studies since the mean flow is also the instantaneous flow because there are no velocity fluctuations present.

Turbulent Boundary Layers

The treatment of turbulent boundary layers is far more difficult due to the time-dependent variation of the flow properties. One of the most widely used techniques in which turbulent flows are tackled is to apply Reynolds decomposition. Here the instantaneous flow properties are decomposed into a mean and fluctuating component. Applying this technique to the boundary layer equations gives the full turbulent boundary layer equations not often given in literature:

$$\frac{\partial \overline{u}}{\partial x} + \frac{\partial \overline{v}}{\partial y} = 0$$

$$\overline{u}\frac{\partial \overline{u}}{\partial x} + \overline{v}\frac{\partial \overline{u}}{\partial y} = -\frac{1}{\rho}\frac{\partial \overline{p}}{\partial x} + \nu\left(\frac{\partial^2 \overline{u}}{\partial x^2} + \frac{\partial^2 \overline{u}}{\partial y^2}\right) - \frac{\partial}{\partial y}(\overline{u'v'}) - \frac{\partial}{\partial x}(\overline{u'^2})$$

$$\overline{u}\frac{\partial \overline{v}}{\partial x} + \overline{v}\frac{\partial \overline{v}}{\partial y} = -\frac{1}{\rho}\frac{\partial \overline{p}}{\partial y} + \nu\left(\frac{\partial^2 \overline{v}}{\partial x^2} + \frac{\partial^2 \overline{v}}{\partial y^2}\right) - \frac{\partial}{\partial x}(\overline{u'v'}) - \frac{\partial}{\partial y}(\overline{v'^2})$$

Using the same order-of-magnitude analysis as for the instantaneous equations, these turbulent boundary layer equations generally reduce to become in their classical form:

$$\frac{\partial \overline{u}}{\partial x} + \frac{\partial \overline{v}}{\partial y} = 0$$

$$\overline{u}\frac{\partial \overline{u}}{\partial x} + \overline{v}\frac{\partial \overline{u}}{\partial y} = -\frac{1}{\rho}\frac{\partial \overline{p}}{\partial x} + \nu\frac{\partial^2 \overline{u}}{\partial y^2} - \frac{\partial}{\partial y}(\overline{u'v'})$$

$$\frac{\partial \overline{p}}{\partial y} = 0$$

The additional term $\overline{u'v'}$ in the turbulent boundary layer equations is known as the Reynolds shear stress and is unknown a priori. The solution of the turbulent boundary layer equations therefore necessitates the use of a turbulence model, which aims to express the Reynolds shear stress in terms of known flow variables or derivatives. The lack of accuracy and generality of such models is a major obstacle in the successful prediction of turbulent flow properties in modern fluid dynamics.

A laminar sub-layer exists in the turbulent zone; it occurs due to those fluid molecules which are still in the very proximity of the surface, where the shear stress is maximum and the velocity of fluid molecules is zero.

Heat and Mass Transfer

In 1928, the French engineer André Lévêque observed that convective heat transfer in a flowing fluid is affected only by the velocity values very close to the surface. For flows of large Prandtl number, the temperature/mass transition from surface to freestream temperature takes place across a very thin region close to the surface. Therefore, the most important fluid velocities are those inside this very thin region in which the change in velocity can be considered linear with normal distance from the surface. In this way, for

$$u(y) = u_0\left[1 - \frac{(y-h)^2}{h^2}\right] = u_0\frac{y}{h}\left[2 - \frac{y}{h}\right],$$

when $y \to 0$, then

$$u(y) \approx 2u_0\frac{y}{h} = \theta y,$$

where θ is the tangent of the Poiseuille parabola intersecting the wall. Although Lévêque's solution was specific to heat transfer into a Poiseuille flow, his insight helped lead other scientists to an exact solution of the thermal boundary-layer problem. Schuh observed that in a boundary-layer, u is again a linear function of y, but that in this case, the wall tangent is a function of x. He expressed this with a modified version of Lévêque's profile,

$$u(y) = \theta(x)y.$$

This results in a very good approximation, even for low Pr numbers, so that only liquid metals with Pr much less than 1 cannot be treated this way. In 1962, Kestin and Persen published a paper describing solutions for heat transfer when the thermal boundary layer is contained entirely within the momentum layer and for various wall temperature distributions. For the problem of a flat plate with a temperature jump at $x = x_0$, they propose a substitution that reduces the parabolic thermal boundary-layer equation to an ordinary differential equation. The solution to this equation, the temperature at any point in the fluid, can be expressed as an incomplete gamma function. Schlichting proposed an equivalent substitution that reduces the thermal boundary-layer equation to an ordinary differential equation whose solution is the same incomplete gamma function.

Convective Transfer Constants from Boundary Layer Analysis

Paul Richard Heinrich Blasius derived an exact solution to the above laminar boundary layer equations. The thickness of the boundary layer is a function of the Reynolds number for laminar flow.

$$\delta \approx \frac{5.0 * x}{\sqrt{Re}}$$

δ = the thickness of the boundary layer: the region of flow where the velocity is less than 99% of the far field velocity v_∞; x is position along the semi-infinite plate, and Re is the Reynolds Number given by $\rho v_\infty x / \mu$ (ρ = density and μ = dynamic viscosity).

The Blasius solution uses boundary conditions in a dimensionless form:

$$\frac{v_x - v_S}{v_\infty - v_S} = \frac{v_x}{v_\infty} = \frac{v_y}{v_\infty} = 0 \quad \text{at} \quad y = 0$$

$$\frac{v_x - v_S}{v_\infty - v_S} = \frac{v_x}{v_\infty} = 1 \quad \text{at} \quad y = \infty \text{ and } x = 0$$

Note that in many cases, the no-slip boundary condition holds that v_S, the fluid velocity at the surface of the plate equals the velocity of the plate at all locations. If the plate is not moving, then $v_S = 0$. A much more complicated derivation is required if fluid slip is allowed.

In fact, the Blasius solution for laminar velocity profile in the boundary layer above a semi-infinite plate can be easily extended to describe Thermal and Concentration boundary layers for heat and mass transfer respectively. Rather than the differential x-momentum balance (equation of motion), this uses a similarly derived Energy and Mass balance:

$$\text{Energy: } v_x \frac{\partial T}{\partial x} + v_y \frac{\partial T}{\partial y} = \frac{k}{\rho Cp} \frac{\partial^2 T}{\partial y^2}$$

$$\text{Mass: } v_x \frac{\partial c_A}{\partial x} + v_y \frac{\partial c_A}{\partial y} = D_{AB} \frac{\partial^2 c_A}{\partial y^2}$$

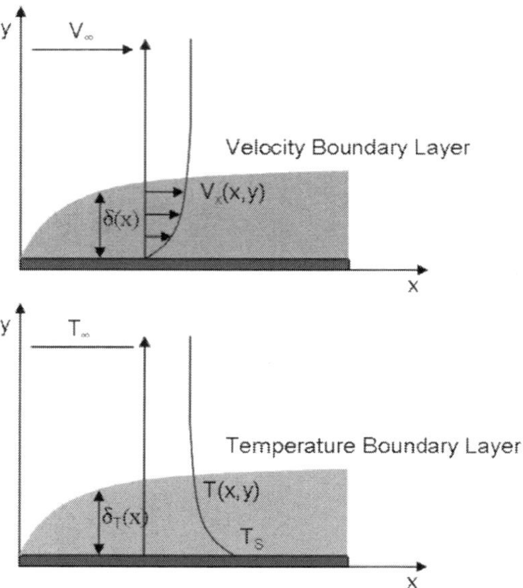

Figure: *Velocity Boundary Layer (Top) and Temperature Boundary Layer (Bottom) share a functional form due to similarity in the Momentum/ Energy Balances and boundary conditions.*

For the momentum balance, kinematic viscosity ν can be considered to be the *momentum diffusivity*. In the energy balance this is replaced by thermal diffusivity $\alpha = k / \rho C_P$, and by mass diffusivity D_{AB} in the mass balance. In thermal diffusivity of a substance, k is its thermal conductivity, ρ is its density and C_p is its heat capacity. Subscript AB denotes diffusivity of species A diffusing into species B.

Under the assumption that $\alpha = D_{AB} = \nu$, these equations become equivalent to the momentum balance. Thus, for Prandtl number $Pr = \nu / \alpha = 1$ and Schmidt number $Sc = \nu / D_{AB} = 1$ the Blasius solution applies directly.

Accordingly, this derivation uses a related form of the boundary conditions, replacing v with T or c_A (absolute temperature or concentration of species A). The subscript S denotes a surface condition.

$$\frac{v_x - v_S}{v_\infty - v_S} = \frac{T - T_S}{T_\infty - T_S} = \frac{c_A - c_{AS}}{c_{A\infty} - c_{AS}} = 0 \text{ at } y = 0$$

$$\frac{v_x - v_S}{v_\infty - v_S} = \frac{T - T_S}{T_\infty - T_S} = \frac{c_A - c_{AS}}{c_{A\infty} - c_{AS}} = 1 \text{ at } y = \infty \text{ and } x = 0$$

Using the streamline function Blasius obtained the following solution for the shear stress at the surface of the plate.

$$\tau_0 = \left(\frac{\partial v_x}{\partial y}\right)_{y=0} = 0.332\frac{v_\infty}{x}Re^{1/2}$$

And via the boundary conditions, it is known that

$$\frac{v_x - v_S}{v_\infty - v_S} = \frac{T - T_S}{T_\infty - T_S} = \frac{c_A - c_{AS}}{c_{A\infty} - c_{AS}}$$

We are given the following relations for heat/mass flux out of the surface of the plate

$$\left(\frac{\partial T}{\partial y}\right)_{y=0} = 0.332\frac{T_\infty - T_S}{x}Re^{1/2}$$

$$\left(\frac{\partial c_A}{\partial y}\right)_{y=0} = 0.332\frac{c_{A\infty} - c_{AS}}{x}Re^{1/2}$$

So for $Pr = Sc = 1$

$$\delta = \delta_T = \delta_c = \frac{5.0 * x}{\sqrt{Re}}$$

Where δ_T, δ_c are the regions of flow where T and c_A are less than 99% of their far field values.

Because the Prandtl number of a particular fluid is not often unity, German engineer E. Polhausen who worked with Ludwig Prandtl attempted to empirically extend these equations to apply for $Pr \neq 1$. His results can be applied to Sc as well. He found that for Prandtl number greater than 0.6, the thermal boundary layer thickness was approximately given by:

$$\frac{\delta}{\delta_T} = Pr^{1/3} \text{ and therefore } \frac{\delta}{\delta_c} = Sc^{1/3}$$

From this solution, it is possible to characterize the convective heat/mass transfer constants based on the region of boundary layer flow. Fourier's law of conduction and Newton's Law of Cooling are combined with the flux term derived above and the boundary layer thickness.

$$\frac{q}{A} = -k\left(\frac{\partial T}{\partial y}\right)_{y=0} = h_x(T_S - T_\infty)$$

$$h_x = 0.332\frac{k}{x}Re_x^{1/2}Pr^{1/3}$$

This gives the local convective constant h_x at one point on the semi-infinite plane.

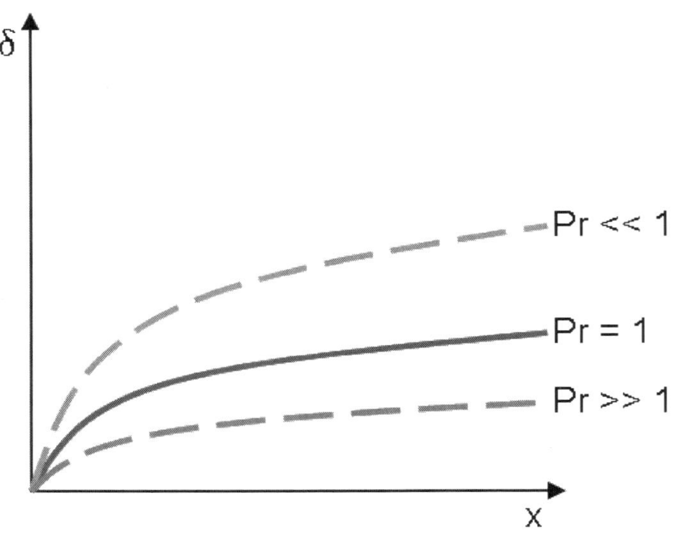

Figure: *Plot showing the relative thickness in the Thermal boundary layer versus the Velocity boundary layer (in red) for various Prandtl Numbers. For* $Pr = 1$*, the two are equal.*

Integrating over the length of the plate gives an average

$$h_L = 0.664\frac{k}{x}Re_L^{1/2}Pr^{1/3}$$

Following the derivation with mass transfer terms (k = convective mass transfer constant, D_{AB} = diffusivity of species A into species B, $Sc = \nu / D_{AB}$), the following solutions are obtained:

$$k'_x = 0.332\frac{D_{AB}}{x}Re_x^{1/2}Sc^{1/3}$$

$$k'_L = 0.664\frac{D_{AB}}{x}Re_L^{1/2}Sc^{1/3}$$

These solutions apply for laminar flow with a Prandtl/Schmidt number greater than 0.6.

Boundary Layer Turbine

This effect was exploited in the Tesla turbine, patented by Nikola Tesla in 1913. It is referred to as a bladeless turbine because it uses the boundary layer effect and not a fluid impinging upon the blades as in a conventional turbine. Boundary layer turbines are also known as cohesion-type turbine, bladeless turbine, and Prandtl layer turbine

Low-Latitude Boundary Layer

The low-latitude boundary layer, first discussed by Eastman et al. (1976), is part of the magnetospheric boundary layer. Large fraction of all the plasma, momentum, and energy transfer from the magnetosheath into the magnetosphere occurs through it.

Characteristics

- low latitude part of the dayside boundary layer extending also to the evening local times
- thickness increases with increasing distance from the subsolar point, being about 0.5-1 Re at the dawn-dusk meridian
- may be partially on open field lines, especially during southward IMF Bz, when it is not as thick as during northward IMF
- mixture of plasma of magnetospheric and magnetosheath origin; the magnetosheath plasma may be leaking diffusively through the magnetopause into this region
- region of general tailward plasma flow
  - o there is some evidence that, in the magnetospheric side of the LLBL, there exists a stagnation region with LLBL-like plasma with variable flow (both in direction and magnitude); this could be due to Kelvin-Helmholtz instability
    - at low altitudes this means that the convection reversal boundary may not coincide exactly with the LLBL/BPS particle boundary, but occurs withing the LLBL
- site of the dayside region 1 current system
  - o most intense upward current close to 1400-1600 MLT, and downward 0800-1000 MLT
- partially thermalized with respect to magnetosheath plasma, with lower flow velocities and higher temperatures (with considerable variability and spatial inhomogeneity, however)
  - o densities: 0.5 - 10 cm^-3

- o temperatures: from less than 100 eV to 1000-2000 eV
- o flow: generally tailward from about 100 km/s up to magnetosheath values (or higher, like 800 km/s)

- region of bidirectional streaming electron populations (50-200 eV)
- there is a systematic relationship between the electron number density and temperature, i.e., the transition parameter.
- The so called "halo" region Earthward of the dayside LLBL may relate to the LLBL itself or the cold plasma sheet.

Low Altitude Parachute Extraction System

Low Altitude Parachute Extraction System (LAPES) is a tactical military airlift delivery method where a fixed wing cargo aircraft can deposit supplies when landing is not an option in an area that is too small to accurately parachute supplies from a high altitude.

This method was developed by the US Military with the assistance of the 109th Quartermaster Company (Air Drop) in 1964. In May 1965, a detachment of the 109th was formed as the 383rd Quartermaster (Aerial Supply) Detachment and sent to Vietnam.

In 1966 the 109th was sent to Vietnam and took operational control of the 383rd. Both units provided Air Drop and LAPES support during the Siege of Khe Sanh in the Vietnam War. LAPES was used to provide a method of supplying heavy loads into Khe Sanh which could not effectively be supplied by air drop. This practice was perfected at Mactan Air Base in Cebu, the Philippines.

LAPES involves loading supplies on a special pallet on a plane. In preparation for a drop the cargo door and ramp of the aircraft is opened and a drogue parachute released. The aircraft descends to the drop altitude (typically only a few metres above the ground). Once the drop plane reaches the desired drop point, the braking parachutes for the load are released and they are extracted from the aircraft by the drogue chute and the drop load is released (retaining straps cut). The braking parachutes pull the load from the aircraft and bring it to a stop on the ground within the drop zone.

The main parachutes are sized to stop the movement of the load sliding on the ground within the required space, and are not intended to control the descent of the load to the ground. Cushioning of the load is accomplished by the pallet and the material between the pallet and the load. Once the delivery is accomplished, the pilot ascends to a normal altitude and returns to base.

LAPES enables planes to quickly deploy large cargo in a timely fashion instead of having to land and take off, which exposes the plane to enemy fire. The technique also allows delivery of loads that are too heavy for a direct parachute descent (high altitude drop). However, the drop sequence's low altitude allows for no margin of pilot error and the risk of plane crash is heightened.

On July 1, 1987 during a Capabilities exercise (CAPEX) a USAF C-130E (68-10945 c/n 4325) crashed while performing a LAPES demo at the Sicily Drop Zone, on Ft. Bragg. The pilot of the C-130 had performed a LAPES drop with the same extreme rate of descent two days earlier during a practice run. The crash killed three on board, one soldier on the ground, and injured two crew.

Effects on Health

All athletes seek a competitive advantage. Although the benefits of some interventions (like training, for example) are clear, most strategies are less well proven. Altitude is no exception to this. Training at high altitude has been used by competitive athletes as a means of improving their potential. However, despite a good deal of research into the topic, its true effects and a recommended approach are still not well established. Additionally, altitude training is usually expensive and fraught with logistical problems.

Benefits of Altitude Exposure

Exposure to high altitude could theoretically improve an athlete's capacity to exercise. Exposing the body to high altitude causes it to acclimatise to the lower level of oxygen available in the atmosphere. Many of the changes that occur with acclimatisation improve the delivery of oxygen to the muscles -the theory being that more oxygen will lead to better performance.

For any type of exercise lasting longer than a few minutes, the body must use oxygen to generate energy. Without it, muscles simply seize up and can become damaged. This type of exercise is called aerobic exercise, meaning with oxygen.

The body naturally produces a hormone called erythropoetin (EPO) which stimulates the production of red blood cells which carry oxygen to the muscles. Up to a point, the more blood cells you have, the more oxygen you can deliver to your muscles. There are also a number of other changes that happen during acclimatisation which may help athletic performance, including an increase in the number of small blood vessels, an increase in buffering capacity (ability to manage the build up of waste acid) and changes in the microscopic structure and function of the muscles themselves.

Problems of Altitude Exposure

However, acclimatisation to high altitude is not simple, and there are a number of other effects that could cancel out the above benefits. For example the increase in red blood cells comes at a cost - having too many blood cells makes the blood thicker and can make blood flow sluggish. This makes it harder for your heart to pump round the body, and can actually decrease the amount of oxygen getting to where it is needed.

At very high altitudes (>5000m), weight loss is unavoidable because your body actually consumes your muscles in order to provide energy. There is even a risk that the body's immune system will become weakened, leading to an increased risk of infections, and there may be adverse changes in the chemical make-up of the muscles. Additionally, the body cannot exercise as intensely at altitude.

This results in reduced training intensity, which can reduce performance in some sports. At very high altitudes, further problems are encountered: loss of appetite, inhibition of muscle repair processes and excessive work of breathing. On top of this, there is the problem of altitude illnesses, which can dramatically reduce the capacity to be active at altitude, or foreshorten the exposure to high altitude altogether.

Altitude Exposure Techniques

Taking some of the information about altitude training into consideration, various techniques have been devised in order to expose the athlete to the beneficial effects of high altitude whilst not reducing their ability to train effectively. These have been labelled 'Live High – Train High, 'Live Low – Train High' and 'Live High – Train Low'.

The typical altitudes used are around 2000-2500m, which in itself reduces the risk of some of the unhelpful effects of altitude exposure. The 'low' altitudes may not actually be at sea level, but could be 1250m, for example. However, the difference between the two altitudes is significant enough to have an effect on training.

Live High – Train High

Maximum exposure to altitude. Evidence of a positive effect at sea level is controversial, and there is less support for this method amongst experts.

Live Low – Train High

The idea behind this regime is that the athlete is exercising in a low oxygen environment, whilst resting in a normal oxygen environment. There have been some interesting findings suggesting that this technique might work, but there are no good studies showing that the technique makes any difference to the ultimate competitive performance of the athlete at sea-level. Additionally, training intensity is reduced so some athletes may find that they actually lose fitness using this regime.

Live High – Train Low

The theory behind this regime is that the body will acclimatise to altitude by living there, whilst training intensity can be maintained by training at (or near) sea level. Hence, the beneficial effects of altitude exposure are harnessed whilst some of the negative ones are avoided. However, residence at altitude must be for more than 12 hours per day and for at least 3 weeks. With this technique, improvements in sea-level performance have been shown in events lasting between 8 and 20 minutes. And interestingly, athletes of all abilities are thought to benefit.

Plasma

Plasma is one of the four fundamental states of matter, the others being solid, liquid, and gas. A plasma has properties unlike those of the other states.

A plasma can be created by heating a gas or subjecting it to a strong electromagnetic field applied with a laser or microwave generator. This reduces or increases the number of electrons, creating positive or negative charged particles called ions, and is accompanied by the dissociation of molecular bonds, if present.

The presence of a non-negligible number of charge carriers makes plasma electrically conductive so that it responds strongly to electromagnetic fields. Like gas, plasma does not have a definite shape or a definite volume unless enclosed in a container. Unlike gas, under the influence of a magnetic field, it may form structures such as filaments, beams and double layers.

Plasma is the most abundant form of ordinary matter in the Universe, most of which is in the rarefied intergalactic regions, particularly the intracluster medium, and in stars, including the Sun. A common form of plasmas on Earth is seen in neon signs.

Much of the understanding of plasmas has come from the pursuit of controlled nuclear fusion and fusion power, for which plasma physics provides the scientific basis.

Definition

Plasma is loosely described as an electrically neutral medium of unbound positive and negative particles (i.e. the overall charge of a plasma is roughly zero). It is important to note that although they are unbound, these particles are not 'free' in the sense of not experiencing forces. When the charges move, they generate electrical currents with magnetic fields, and as a result, they are affected by each other's fields. This governs their collective behaviour with many degrees of freedom. A definition can have three criteria:

1. The plasma approximation: Charged particles must be close enough together that each particle influences many nearby charged particles, rather than just interacting with the closest particle (these collective effects are a distinguishing feature of a plasma). The plasma approximation is valid when the number of charge carriers within the sphere of influence (called the *Debye sphere* whose radius is the Debye screening length) of a particular particle is higher than unity to provide collective behaviour of the charged particles. The average number of particles in the Debye sphere is given by the plasma parameter, "Ë" (the Greek letter Lambda).
2. Bulk interactions: The Debye screening length (defined above) is short compared to the physical size of the plasma. This criterion means that interactions in the bulk of the plasma are more important than those at its edges, where boundary effects may take place. When this criterion is satisfied, the plasma is quasineutral.
3. Plasma frequency: The electron plasma frequency (measuring plasma oscillations of the electrons) is large compared to the electron-neutral collision frequency (measuring frequency of collisions between electrons and neutral particles). When this condition is valid, electrostatic interactions dominate over the processes of ordinary gas kinetics.

Ranges of Parameters

Plasma parameters can take on values varying by many orders of magnitude, but the properties of plasmas with apparently disparate parameters may be very similar. The following chart considers only conventional atomic plasmas and not exotic phenomena like quark gluon plasmas:

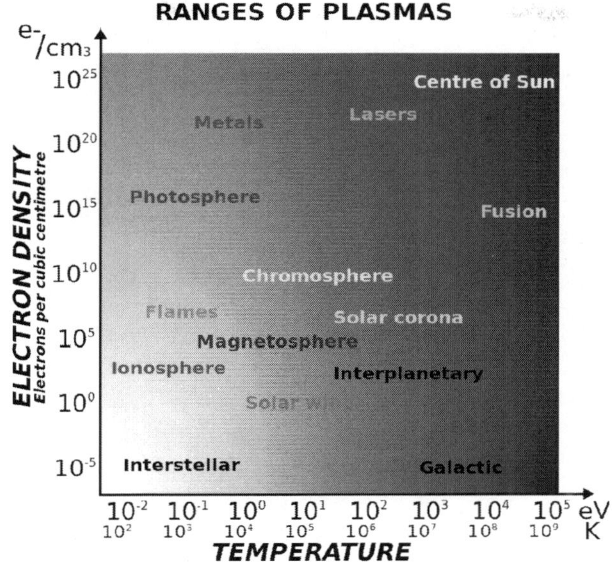

Figure: *Range of plasmas. Density increases upwards, temperature increases towards the right. The free electrons in a metal may be considered an electron plasma.*

Typical ranges of plasma parameters: orders of magnitude (OOM)		
Characteristic	***Terrestrial plasmas***	***Cosmic plasmas***
Size in meters	10^{-6} m (lab plasmas) to 10^{2} m (lightning) (~8 OOM)	10^{-6} m (spacecraft sheath) to 10^{25} m (intergalactic nebula) (~31 OOM)
Lifetime in seconds	10^{-12} s (laser-produced plasma) to 10^{7} s (fluorescent lights) (~19 OOM)	10^{1} s (solar flares) to 10^{17} s (intergalactic plasma) (~16 OOM)
Density in particles per cubic meter	10^{7} m^{-3} to 10^{32} m^{-3} (inertial confinement plasma)	1 m^{-3} (intergalactic medium) to 10^{30} m^{-3} (stellar core)
Temperature in Kelvin	~0 K (crystalline non-neutral plasma) to 10^{8} K (magnetic fusion plasma)	10^{2} K (aurora) to 10^{7} K (solar core)
Magnetic fields in teslas	10^{-4} T (lab plasma) to 10^{3} T (pulsed-power plasma)	10^{-12} T (intergalactic medium) to 10^{11} T (near neutron stars)

Degree of Ionization

For plasma to exist, ionization is necessary. The term "plasma density" by itself usually refers to the "electron density", that is, the number of free electrons per unit volume. The degree of ionization of a plasma is the proportion of atoms that have lost or gained electrons, and is controlled mostly by the temperature. Even a partially ionized gas in which as little as 1% of the particles are ionized can have the characteristics of a plasma (i.e., response to magnetic fields and high electrical conductivity). The degree of ionization, α, is defined as $\alpha = \frac{n_i}{n_i + n_n}$, where n_i is the number density of ions and n_n is the number density of neutral atoms. The *electron density* is related to this by the average charge state $\langle Z \rangle$ of the ions through $n_e = \langle Z \rangle n_i$, where n_e is the number density of electrons.

Temperatures

Plasma temperature is commonly measured in Kelvins or electronvolts and is, informally, a measure of the thermal kinetic energy per particle. Very high temperatures are usually needed to sustain ionization, which is a defining feature of a plasma. The degree of plasma ionization is determined by the "electron temperature" relative to the ionization energy (and more weakly by the density), in a relationship called the Saha equation. At low temperatures, ions and electrons tend to recombine into bound states—atoms—and the plasma will eventually become a gas.

In most cases the electrons are close enough to thermal equilibrium that their temperature is relatively well-defined, even when there is a significant deviation from a Maxwellian energy distribution function, for example, due to UV radiation, energetic particles, or strong electric fields. Because of the large difference in mass, the electrons come to thermodynamic equilibrium amongst themselves much faster than they come into equilibrium with the ions or neutral atoms. For this reason, the "ion temperature" may be very different from (usually lower than) the "electron temperature". This is especially common in weakly ionized technological plasmas, where the ions are often near the ambient temperature.

Nonthermal Plasma

A nonthermal plasma is in general any plasma which is not in thermodynamic equilibrium, either because the ion temperature is different from the electron temperature, or because the velocity distribution of one of the species does not follow a Maxwell–Boltzmann distribution.

In the context of food processing, a nonthermal plasma (NTP) is specifically an antimicrobial treatment being investigated for application to fruits, vegetables and other foods with fragile surfaces. These foods are either not adequately sanitized or are otherwise unsuitable for treatment with chemicals, heat or other conventional food processing tools. The term *cold plasma* has been recently used as a convenient descriptor to distinguish the one-atmosphere, near room temperature plasma discharges from other plasmas, operating at hundreds or thousands of degrees above ambient. Within the context of food processing the term "cold" can potentially engender misleading images of refrigeration requirements as a part of the plasma treatment. However, in practice this confusion has not been an issue.

Nonthermal plasma also sees increasing use in the sterilization of teeth and hands as well as in self-decontaminating filters.

Thermal vs. Non-Thermal Plasmas

Based on the relative temperatures of the electrons, ions and neutrals, plasmas are classified as "thermal" or "non-thermal". Thermal plasmas have electrons and the heavy particles at the same temperature, i.e., they are in thermal equilibrium with each other. Non-thermal plasmas on the other hand have the ions and neutrals at a much lower temperature (sometimes room temperature), whereas electrons are much "hotter" ($T_e \gg T_n$).

A plasma is sometimes referred to as being "hot" if it is nearly fully ionized, or "cold" if only a small fraction (for example 1%) of the gas molecules are ionized, but other definitions of the terms "hot plasma" and "cold plasma" are common. Even in a "cold" plasma, the electron temperature is still typically several thousand degrees Celsius. Plasmas utilised in "plasma technology" ("technological plasmas") are usually cold plasmas in the sense that only a small fraction of the gas molecules are ionized.

Plasma Potential

Since plasmas are very good electrical conductors, electric potentials play an important role. The potential as it exists on average in the space between charged particles, independent of the question of how it can be measured, is called the "plasma potential", or the "space potential". If an electrode is inserted into a plasma, its potential will generally lie considerably below the plasma potential due to what is termed a Debye sheath. The good electrical conductivity of plasmas makes their electric fields very small. This results in the important concept of "quasineutrality", which says the density of negative charges is approximately equal to the density of positive charges over large volumes of the plasma ($n_e = \langle Z \rangle n_i$), but on the scale of the Debye length there can be charge imbalance. In the special case that *double layers* are formed, the charge separation can extend some tens of Debye lengths.

The magnitude of the potentials and electric fields must be determined by means other than simply finding the net charge density. A common example is to assume that the electrons satisfy the Boltzmann relation:

$$n_e \propto e^{e\Phi/k_B T_e}.$$

Differentiating this relation provides a means to calculate the electric field from the density:

$$\vec{E} = (k_B T_e / e)(\nabla n_e / n_e).$$

It is possible to produce a plasma that is not quasineutral. An electron beam, for example, has only negative charges. The density of a non-neutral plasma must generally be very low, or it must be very small, otherwise it will be dissipated by the repulsive electrostatic force.

In astrophysical plasmas, Debye screening prevents electric fields from directly affecting the plasma over large distances, i.e., greater than the Debye length. However, the existence of charged particles causes the plasma to generate, and be affected by, magnetic fields. This can and does cause extremely complex behaviour, such as the generation of plasma double layers, an object that separates charge over a few tens of Debye lengths. The dynamics of plasmas interacting with external and self-generated magnetic fields are studied in the academic discipline of magnetohydrodynamics.

Magnetization

Plasma with a magnetic field strong enough to influence the motion of the charged particles is said to be magnetized. A common quantitative criterion is that a particle on average completes at least one gyration around the magnetic field before making a collision, i.e., $\omega_{\mathrm{ce}} / v_{\mathrm{coll}} > 1$, where ω_{ce} is the "electron gyrofrequency" and v_{coll} is the "electron collision rate". It is often the case that the electrons are magnetized while the ions are not. Magnetized plasmas are *anisotropic*, meaning that their properties in the direction parallel to the magnetic field are different from those perpendicular to it. While electric fields in plasmas are usually small due to the high conductivity, the electric field associated with a plasma moving in a magnetic field is given by $\mathrm{E} = -v \times \mathrm{B}$ (where E is the electric field, v is the velocity, and B is the magnetic field), and is not affected by Debye shielding.

Comparison of Plasma and Gas Phases

Plasma is often called the *fourth state of matter* after solid, liquids and gases. It is distinct from these and other lower-energy states of matter. Although it is closely related to the gas phase in that it also has no definite form or volume, it differs in a number of ways, including the following:

Property	*Gas*	*Plasma*
Electrical conductivity	Very low: Air is an excellent insulator until it breaks down into plasma at electric field strengths above 30 kilovolts per centimeter.	Usually very high: For many purposes, the conductivity of a plasma may be treated as infinite.
Independently acting species	One: All gas particles behave in a similar way, influenced by gravity and by collisions with one another.	Two or three: Electrons, ions, protons and neutrons can be distinguished by the sign and value of their charge so that they behave independently in many circumstances, with different bulk velocities and temperatures, allowing phenomena such as new types of waves and instabilities.
Velocity distribution	Maxwellian: Collisions usually lead to a Maxwellian velocity distribution of all gas particles, with very few relatively fast particles.	Often non-Maxwellian: Collisional interactions are often weak in hot plasmas and external forcing can drive the plasma far from local equilibrium and lead to a significant population of unusually fast particles.
Interactions	Binary: Two-particle collisions are the rule, three-body collisions extremely rare.	Collective: Waves, or organized motion of plasma, are very important because the particles can interact at long ranges through the electric and magnetic forces.

Common Plasmas

Plasmas are by far the most common phase of ordinary matter in the universe, both by mass and by volume. Essentially, all of the visible light from space comes from stars, which are plasmas with a temperature such that they radiate strongly at visible wavelengths. Most of the ordinary (or baryonic) matter in the universe, however, is found in the intergalactic medium, which is also a plasma, but much hotter, so that it radiates primarily as X-rays.

In 1937, Hannes Alfvén argued that if plasma pervaded the universe, it could then carry electric currents capable of generating a galactic magnetic field. After winning the Nobel Prize, he emphasized that:

In order to understand the phenomena in a certain plasma region, it is necessary to map not only the magnetic but also the electric field and the electric currents. Space is filled with a network of currents which transfer energy and momentum over large or very large distances. The currents often pinch to filamentary or surface currents. The latter are likely to give space, as also interstellar and intergalactic space, a cellular structure.

By contrast the current scientific consensus is that about 96% of the total energy density in the universe is not plasma or any other form of ordinary matter, but a combination of cold dark matter and dark energy. Our Sun, and all stars, are made of plasma, much of interstellar space is filled with a plasma, albeit a very sparse one, and intergalactic space too. Even black holes, which are not directly visible, are thought to be fuelled by accreting ionising matter (i.e. plasma), and they are associated with astrophysical jets of luminous ejected plasma, such as M87's jet that extends 5,000 light-years.

In our solar system, interplanetary space is filled with the plasma of the Solar Wind that extends from the Sun out to the heliopause. However, the density of ordinary matter is much higher than average and much higher than that of either dark matter or dark energy. The planet Jupiter accounts for most of the *non*-plasma, only about 0.1% of the mass and 10% of the volume within the orbit of Pluto.

Dust and small grains within a plasma will also pick up a net negative charge, so that they in turn may act like a very heavy negative ion component of the plasma.

Common Forms of Plasma

Artificially Produced

- Those found in plasma displays, including TVs
- Inside fluorescent lamps (low energy lighting), neon signs
- Rocket exhaust and ion thrusters
- The area in front of a spacecraft's heat shield during re-entry into the atmosphere
- Inside a corona discharge ozone generator
- Fusion energy research
- The electric arc in an arc lamp, an arc welder or plasma torch
- Plasma ball (sometimes called a plasma sphere or plasma globe)
- Arcs produced by Tesla coils (resonant air core transformer or disruptor coil that produces arcs similar to lightning, but with alternating current rather than static electricity)
- Plasmas used in semiconductor device fabrication including reactive-ion etching, sputtering, surface cleaning and plasma-enhanced chemical vapor deposition
- Laser-produced plasmas (LPP), found when high power lasers interact with materials.

- Inductively coupled plasmas (ICP), formed typically in argon gas for optical emission spectroscopy or mass spectrometry
- Magnetically induced plasmas (MIP), typically produced using microwaves as a resonant coupling method
- Static electric sparks

Terrestrial Plasmas

- Lightning
- St. Elmo's fire
- Upper-atmospheric lightning (e.g. Blue jets, Blue starters, Gigantic jets, ELVES)
- Sprites
- The ionosphere
- The plasmasphere
- The polar aurorae
- Some flames
- The polar wind, a plasma fountain

Space and Astrophysical Plasmas

- The Sun and other stars (plasmas heated by nuclear fusion)
- The solar wind
- The interplanetary medium (space between planets)
- The interstellar medium (space between star systems)
- The Intergalactic medium (space between galaxies)
- The Io-Jupiter flux tube
- Accretion discs
- Interstellar nebulae Cometary ion tail

Complex Plasma Phenomena

Although the underlying equations governing plasmas are relatively simple, plasma behaviour is extraordinarily varied and subtle: the emergence of unexpected behaviour from a simple model is a typical feature of a complex system. Such systems lie in some sense on the boundary between ordered and disordered behaviour and cannot typically be described either by simple, smooth, mathematical functions, or by pure randomness. The spontaneous formation of interesting spatial features on a wide range of length scales is one manifestation of plasma complexity. The features are interesting, for example, because they are very sharp, spatially intermittent (the distance between features is much larger than the features themselves), or have a fractal form. Many of these features were first studied in the laboratory, and have subsequently been recognised throughout the universe. Examples of complexity and complex structures in plasmas include:

Filamentation

Striations or string-like structures, also known as birkeland currents, are seen in many plasmas, like the plasma ball, the aurora, lightning, electric arcs, solar flares, and supernova remnants. They are sometimes associated with larger current densities, and the interaction with the magnetic field can form a magnetic rope structure. High power microwave breakdown at atmospheric pressure also leads to the formation of filamentary structures.

Filamentation also refers to the self-focusing of a high power laser pulse. At high powers, the nonlinear part of the index of refraction becomes important and causes a higher index of refraction in the center of the laser beam, where the laser is brighter than at the edges, causing a feedback that focuses the laser even more. The tighter focused laser has a higher peak brightness (irradiance) that forms a plasma. The plasma has an index of refraction lower than one, and causes a defocusing of the laser beam. The interplay of the focusing index of refraction, and the defocusing plasma makes the formation of a long filament of plasma that can be micrometers to kilometers in length. One interesting aspect of the filamentation generated plasma is the relatively low ion density due to defocusing effects of the ionized electrons.

Shocks or Double Layers

Plasma properties change rapidly (within a few Debye lengths) across a two-dimensional sheet in the presence of a (moving) shock or (stationary) double layer. Double layers involve localized charge separation, which causes a large potential difference across the layer, but does not generate an electric field outside the layer. Double layers separate adjacent plasma regions with different physical characteristics, and are often found in current carrying plasmas. They accelerate both ions and electrons.

Electric Fields and Circuits

Quasineutrality of a plasma requires that plasma currents close on themselves in electric circuits. Such circuits follow Kirchhoff's circuit laws and possess a resistance and inductance. These circuits must generally be treated as a strongly coupled system, with the behaviour in each plasma region dependent on the entire circuit. It is this strong coupling between system elements, together with nonlinearity, which may lead to complex behaviour. Electrical circuits in plasmas store inductive (magnetic) energy, and should the circuit be disrupted, for example, by a plasma instability, the inductive energy will be released as plasma heating and acceleration. This is a common explanation for the heating that takes place in the solar corona. Electric currents, and in particular, magnetic-field-aligned electric currents (which are sometimes generically referred to as "Birkeland currents"), are also observed in the Earth's aurora, and in plasma filaments.

Cellular Structure

Narrow sheets with sharp gradients may separate regions with different properties such as magnetization, density and temperature, resulting in cell-like regions. Examples include the magnetosphere, heliosphere, and heliospheric current sheet. Hannes Alfvén wrote: "From the cosmological point of view, the most important new space research discovery is probably the cellular structure of space. As has been seen in every region of space accessible to in situ measurements, there are a number of 'cell walls', sheets of electric currents, which divide space into compartments with different magnetization, temperature, density, etc."

Critical Ionization Velocity

The critical ionization velocity is the relative velocity between an ionized plasma and a neutral gas, above which a runaway ionization process takes place. The critical ionization process is a quite general mechanism for the conversion of the kinetic energy of a rapidly streaming gas into ionization and plasma thermal energy. Critical phenomena in general are typical of complex systems, and may lead to sharp spatial or temporal features.

Ultracold Plasma

Ultracold plasmas are created in a magneto-optical trap (MOT) by trapping and cooling neutral atoms, to temperatures of 1 mK or lower, and then using another laser to ionize the atoms by giving each of the outermost electrons just enough energy to escape the electrical attraction of its parent ion.

One advantage of ultracold plasmas are their well characterized and tunable initial conditions, including their size and electron temperature. By adjusting the wavelength of the ionizing laser, the kinetic energy of the liberated electrons can be tuned as low as 0.1 K, a limit set by the frequency bandwidth of the laser pulse. The ions inherit the millikelvin temperatures of the neutral atoms, but are quickly heated through a process known as disorder induced heating (DIH). This type of non-equilibrium ultracold plasma evolves rapidly, and displays many other interesting phenomena.

One of the metastable states of a strongly nonideal plasma is Rydberg matter, which forms upon condensation of excited atoms.

Non-Neutral Plasma

The strength and range of the electric force and the good conductivity of plasmas usually ensure that the densities of positive and negative charges in any sizeable region are equal ("quasineutrality"). A plasma with a significant excess of charge density, or, in the extreme case, is composed of a single species, is called a non-neutral plasma. In such a plasma, electric fields play a dominant role. Examples are charged particle beams, an electron cloud in a Penning trap and positron plasmas.

Dusty Plasma and Grain Plasma

A dusty plasma contains tiny charged particles of dust (typically found in space). The dust particles acquire high charges and interact with each other. A plasma that contains larger particles is called grain plasma. Under laboratory conditions, dusty plasmas are also called *complex plasmas*.

Impermeable Plasma

Impermeable plasma is a type of thermal plasma which acts like an impermeable solid with respect to gas or cold plasma and can be physically pushed. Interaction of cold gas and thermal plasma was briefly studied by a group led by Hannes Alfvén in 1960s and 1970s for its possible applications in insulation of fusion plasma from the reactor walls. However later it was found that the external magnetic fields in this configuration could induce kink instabilities in the plasma and subsequently lead to an unexpectedly high heat loss to the walls.

In 2013, a group of materials scientists reported that they have successfully generated stable impermeable plasma with no magnetic confinement using only an ultrahigh-pressure blanket of cold gas. While spectroscopic data on the characteristics of plasma were claimed to be difficult to obtain due to the high-pressure, the passive effect of plasma on synthesis of different nanostructures clearly suggested the effective confinement. They also showed that upon maintaining the impermeability for a few tens of seconds, screening of ions at the plasma-gas interface could give rise to a strong secondary mode of heating (known as viscous heating) leading to different kinetics of reactions and formation of complex nanomaterials.

Mathematical Descriptions

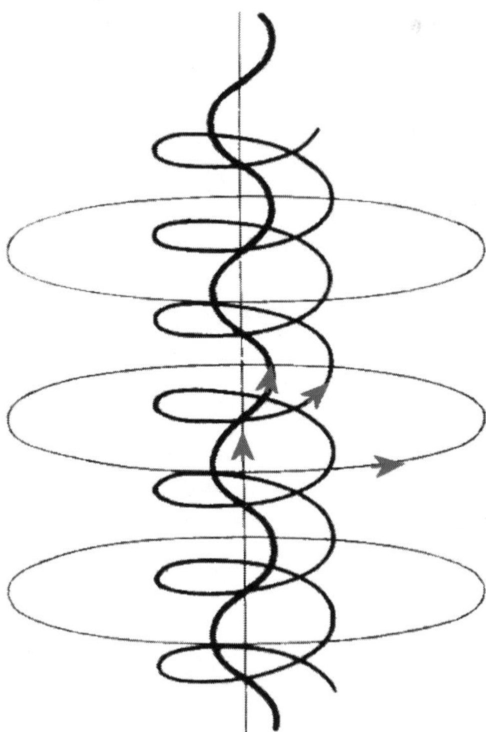

Figure: *The complex self-constricting magnetic field lines and current paths in a field-aligned Birkeland current that can develop in a plasma.*

To completely describe the state of a plasma, we would need to write down all the particle locations and velocities and describe the electromagnetic field in the plasma region. However, it is generally not practical or necessary to keep track of all the particles in a plasma. Therefore, plasma physicists commonly use less detailed descriptions, of which there are two main types:

Fluid Model

Fluid models describe plasmas in terms of smoothed quantities, like density and averaged velocity around each position. One simple fluid model, magnetohydrodynamics, treats the plasma as a single fluid governed by a combination of Maxwell's equations and the Navier–Stokes equations. A more general description is the two-fluid plasma picture, where the ions and electrons are described separately. Fluid models are often accurate when collisionality is sufficiently high to keep the plasma velocity distribution close to a Maxwell–Boltzmann distribution. Because fluid models usually describe the plasma in terms of a single flow at a certain temperature at each spatial location, they can neither capture velocity space structures like beams or double layers, nor resolve wave-particle effects.

Kinetic Model

Kinetic models describe the particle velocity distribution function at each point in the plasma and therefore do not need to assume a Maxwell–Boltzmann distribution. A kinetic description is often necessary for collisionless plasmas. There are two common approaches to kinetic description of a plasma. One is based on representing the smoothed distribution function on a grid in velocity and position. The other, known as the particle-in-cell (PIC) technique, includes kinetic information by following the trajectories of a large number of individual particles. Kinetic models are generally more computationally intensive than fluid models. The Vlasov equation may be used to describe the dynamics of a system of charged particles interacting with an electromagnetic field. In magnetized plasmas, a gyrokinetic approach can substantially reduce the computational expense of a fully kinetic simulation.

Artificial Plasmas

Most artificial plasmas are generated by the application of electric and/or magnetic fields. Plasma generated in a laboratory setting and for industrial use can be generally categorized by:

- The type of power source used to generate the plasma—DC, RF and microwave
- The pressure they operate at—vacuum pressure (< 10 mTorr or 1 Pa), moderate pressure (~ 1 Torr or 100 Pa), atmospheric pressure (760 Torr or 100 kPa)
- The degree of ionization within the plasma—fully, partially, or weakly ionized
- The temperature relationships within the plasma—thermal plasma ($T_e = T_i = T_{gas}$), non-thermal or "cold" plasma ($T_e \gg T_i = T_{gas}$)
- The electrode configuration used to generate the plasma
- The magnetization of the particles within the plasma—magnetized (both ion and electrons are trapped in Larmor orbits

by the magnetic field), partially magnetized (the electrons but not the ions are trapped by the magnetic field), non-magnetized (the magnetic field is too weak to trap the particles in orbits but may generate Lorentz forces)

- The application.

Generation of Artificial Plasma

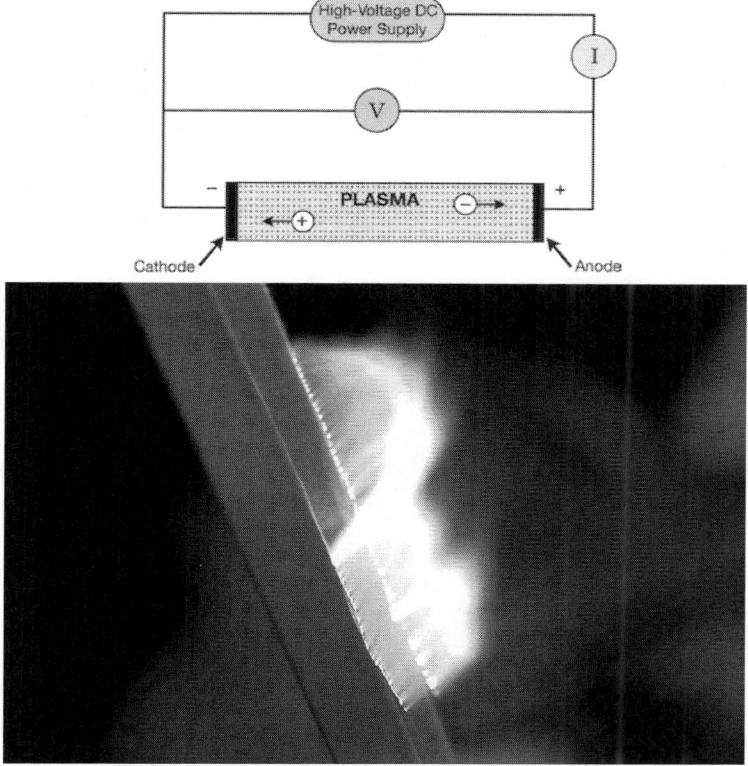

Figure: *Artificial plasma produced in air by a Jacob's Ladder*

Just like the many uses of plasma, there are several means for its generation, however, one principle is common to all of them: there must be energy input to produce and sustain it. For this case, plasma is generated when an electrical current is applied across a dielectric gas or fluid (an electrically non-conducting material) as can be seen in the image to the right, which shows a discharge tube as a simple example (DC used for simplicity).

The potential difference and subsequent electric field pull the bound electrons (negative) toward the anode (positive electrode) while the cathode (negative electrode) pulls the nucleus. As the voltage increases, the current stresses the material (by electric polarization) beyond its dielectric limit (termed strength) into a stage of electrical breakdown, marked by an electric spark, where the material transforms from being an insulator into a conductor (as it becomes increasingly ionized). The underlying process is the Townsend avalanche, where collisions between electrons and neutral gas atoms create more ions and electrons (as can be seen in the figure on the right). The first impact of an electron on an atom results in one ion and two electrons. Therefore, the number of charged particles increases rapidly (in the millions) only "after about 20 successive sets of collisions", mainly due to a small mean free path (average distance travelled between collisions).

Electric Arc

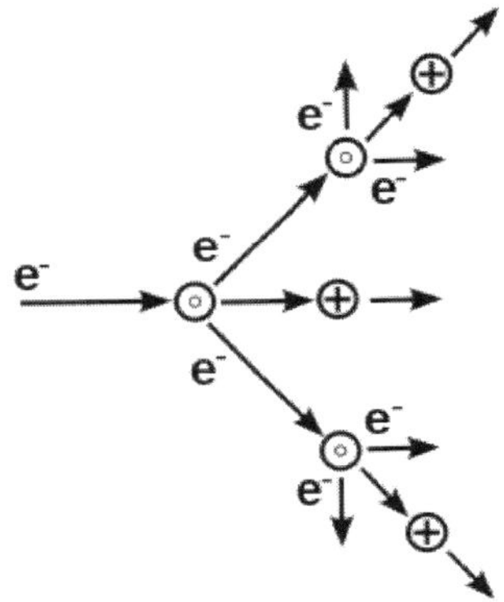

Figure: *Cascade process of ionization. Electrons are 'e⁻', neutral atoms 'o', and cations '+'.*

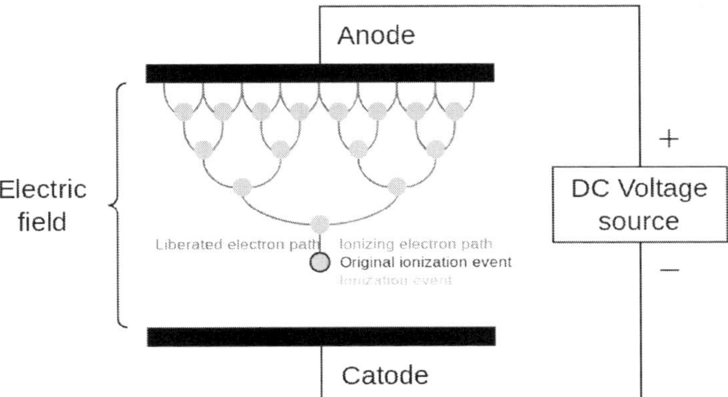

Figure: *Avalanche effect between two electrodes. The original ionisation event liberates one electron, and each subsequent collision liberates a further electron, so two electrons emerge from each collision: the ionising electron and the liberated electron.*

With ample current density and ionization, this forms a luminous electric arc (a continuous electric discharge similar to lightning) between the electrodes. Electrical resistance along the continuous electric arc creates heat, which dissociates more gas molecules and ionizes the resulting atoms (where degree of ionization is determined by temperature), and as per the sequence: solid-liquid-gas-plasma, the gas is gradually turned into a thermal plasma.

A thermal plasma is in thermal equilibrium, which is to say that the temperature is relatively homogeneous throughout the heavy particles (i.e. atoms, molecules and ions) and electrons. This is so because when thermal plasmas are generated, electrical energy is given to electrons, which, due to their great mobility and large numbers, are able to disperse it rapidly and by elastic collision (without energy loss) to the heavy particles.

Examples of Industrial/Commercial Plasma

Because of their sizable temperature and density ranges, plasmas find applications in many fields of research, technology and industry. For example, in: industrial and extractive metallurgy, surface treatments such as plasma spraying (coating), etching in microelectronics, metal cutting and welding; as well as in everyday vehicle exhaust cleanup and fluorescent/luminescent lamps, while even playing a part in supersonic combustion engines for aerospace engineering.

Low-Pressure Discharges

- *Glow discharge plasmas*: non-thermal plasmas generated by the application of DC or low frequency RF (<100 kHz) electric field to the gap between two metal electrodes. Probably the most common plasma; this is the type of plasma generated within fluorescent light tubes.
- *Capacitively coupled plasma (CCP)*: similar to glow discharge plasmas, but generated with high frequency RF electric fields, typically 13.56 MHz. These differ from glow discharges in that the sheaths are much less intense. These are widely used in the microfabrication and integrated circuit manufacturing industries for plasma etching and plasma enhanced chemical vapor deposition.
- *Cascaded Arc Plasma Source*: a device to produce low temperature (~1eV) high density plasmas (HDP).
- *Inductively coupled plasma (ICP)*: similar to a CCP and with similar applications but the electrode consists of a coil wrapped around the chamber where plasma is formed.

- *Wave heated plasma*: similar to CCP and ICP in that it is typically RF (or microwave). Examples include helicon discharge and electron cyclotron resonance (ECR).

Atmospheric Pressure

- *Arc discharge:* this is a high power thermal discharge of very high temperature (~10,000 K). It can be generated using various power supplies. It is commonly used in metallurgical processes. For example, it is used to smelt minerals containing Al_2O_3 to produce aluminium.
- *Corona discharge:* this is a non-thermal discharge generated by the application of high voltage to sharp electrode tips. It is commonly used in ozone generators and particle precipitators.
- *Dielectric barrier discharge (DBD):* this is a non-thermal discharge generated by the application of high voltages across small gaps wherein a non-conducting coating prevents the transition of the plasma discharge into an arc. It is often mislabelled 'Corona' discharge in industry and has similar application to corona discharges. It is also widely used in the web treatment of fabrics. The application of the discharge to synthetic fabrics and plastics functionalizes the surface and allows for paints, glues and similar materials to adhere.
- *Capacitive discharge:* this is a nonthermal plasma generated by the application of RF power (e.g., 13.56 MHz) to one powered electrode, with a grounded electrode held at a small separation distance on the order of 1 cm. Such discharges are commonly stabilized using a noble gas such as helium or argon.

Chapter 6

Usefulness of Geomagnetic Research

Significant contribution to research in geomagnetism started from India as back as in 19th century with the pioneering work of Brown and Chambers and Moos. Institute started work in the areas of solar and lunar geomagnetic variations, equatorial electrojet studies, annual and semi-annual variations of the geomagnetic field, complexities of the recurrent geomagnetic field signatures etc. Solar-terrestrial-physics associated studies were mainly utilising long series of geomagnetic field observations at the Indian Observatories and also worldwide network of geomagnetic data. The Geomagnetic Observatory data were also used for studies on Solar wind plasma and Interplanetary Magnetic Field (IMF) associations, Solar flare effects etc., thus understanding the Sun-Earth interaction phenomena as inferred from the ground magnetic variations over the equatorial and low latitude locations.

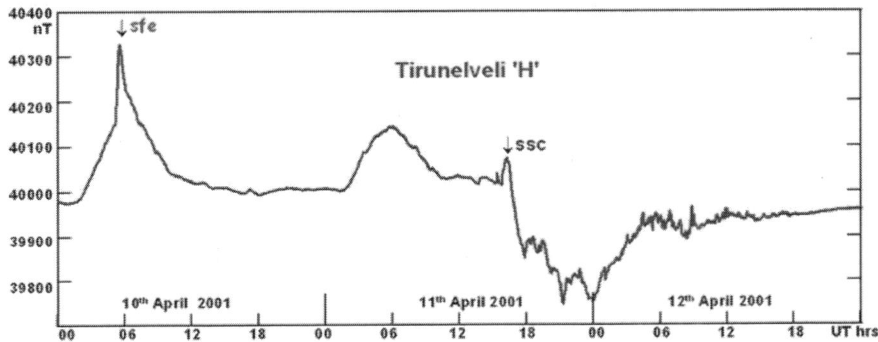

Digital magnetic records at Tirunelveli for three consecutive days during different solar activity conditions in April 2001. Solar flare effect is seen as a spurt in the amplitude of the Horizontal component ('H') on April 10th, Normal day (April 11th) and an intense storm development on April 11th . The impact of the shock associated with the coronal mass ejection following the solar flare on April 10th, is seen as storm Sudden commencement and subsequent development of an intense main phase in the ground magnetic field.

The continuous recorded data gives an opportunity to decipher the long-term secular changes as well as the daily variations of the magnetic components that is basically the reflection of the ionospheric and magnetospheric change occurring above a particular observatory. Thus the variations in the geomagnetic field can be used as a diagnostic tool for understanding the internal structure of the Earth as well as the dynamics of the upper atmosphere and magnetosphere.

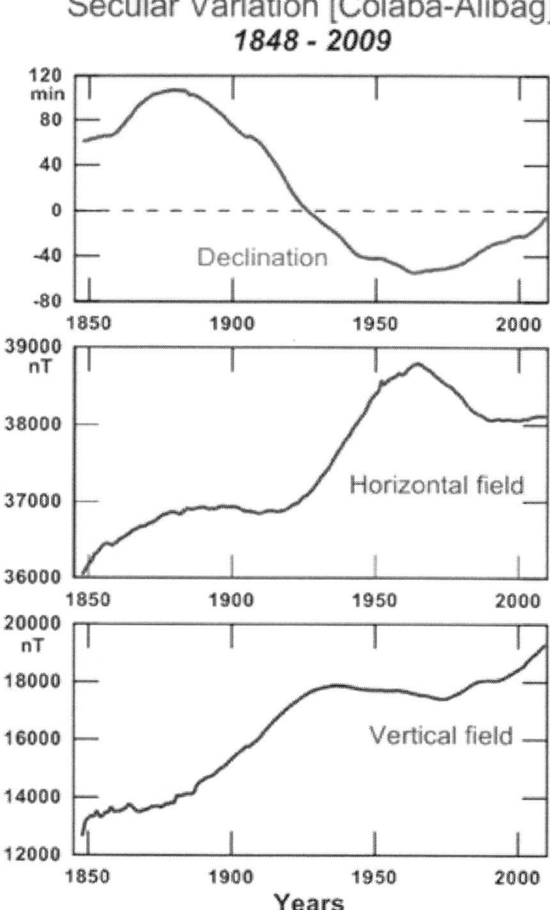

Secular variation of the three components of the earth's magnetic field from the Colaba-Alibag combined observations for 160 years The geographical location of India plays a pivotal role with the latitudinal coverage existing from equator to the focus of the low latitude Sq current system. India has the long lasting history of measuring the earth's magnetic field using classical instruments. With the advancement of instrumentation in the field of magnetic measurements, the observatories operated by IIG are being modernised using fluxgate magnetometers for variation recordings and Declination Inclination Magnetometer (DIM) for Absolute observations.

Solar Quiet (Sq) and Equatorial Electrojet Variations

Locations of the Indian permanent observatories are such as to cover a wide range of latitudes extending from the dip equator up to the focus of the Sq current system, which flows in the E-region of the ionosphere. Thus, the magnetic data recorded continuously at these observatories provides a unique opportunity to study the associated phenomena of ionospheric Sq current system as well as the equatorial electrojet (EEJ).

The EEJ flows eastward within a narrow region of about ? 50 latitude centered over the dip equator. Scientists at the Institute have been working towards understanding the detailed nature of these current systems, including its large day-to-day variability, solar activity and seasonal dependence, as well as the reverse equatorial electrojet, better known as equatorial counter electrojet (CEJ) phenomenon. An empirical model of the equatorial electrojet has been successfully set up based on the surface magnetic data recorded at various stations located in six different longitude sectors, and able to reproduce the characteristic signatures of the EEJ-associated horizontal and vertical magnetic field components at ground level. Besides this, researchers at the institute are also engaged in studying ionospheric electrodynamics using various satellite magnetic field observations such as Magsat, Oersted, Champ, SAC-C etc. The ground magnetic field data from Indian stations have been used to support the satellite findings.

Effects of Magnetic Disturbances

The importance of geomagnetic data in exploring the space weather phenomena of varied timescales is becoming crucial.

The ionosphere and magnetosphere are a closely coupled system that channels energy and momentum from the solar wind to the upper atmosphere. A number of coupled current systems flow in these regions of highly conducting plasmas. These currents are responsible for most of the temporal changes in the geomagnetic field that occur on time scales of seconds to days, including magnetic pulsations. Studies of the ionosphere and magnetosphere seek to obtain a quantitative understanding of the flow of energy and momentum through the solar wind, magnetosphere and ionosphere systems. This study in turn supports the physics of magnetic reconnection at the magnetopause, the response of the magnetosphere to changes in solar wind pressure, the processes responsible for viscous like interactions. Investigations towards this direction elucidate the physical mechanisms responsible for generating pulsations and controlling their cross-field transport in the magnetosphere.

Geomagnetic Field Fluctuations of Various Periodicities

Study of micropulsations, planetary waves, and long period oscillations such as quasi-biennial (QBO), annual (AO), and semi-annual (SAO) oscillations, is being carried out in the institute using high resolution magnetometers.

IIG scientists have attempted to establish the spatial and frequency characteristics of the equatorial enhancement for several period bands from minutes range to sq harmonics using geomagnetic data from upgraded/new networks. Using data adaptive singular spectrum analysis technique, the regular Sq harmonics and Pc pulsations can be isolated. The characteristics of geomagnetic pulsations undergo appreciable changes as they pass through the ionosphere. These changed properties at the low and equatorial stations are distinctly different from those at the high latitudes. It is found that polarization directions of PC3-4 (period 10 to 100 seconds) pulsation changed during the counter electrojet time. The amplitude of these pulsations is enhanced by equatorial electrojet.

Geomagnetic pulsation in PC5-6 range is being studied using the magnetic records. It is seen that these pulsations are enhanced in the equatorial region by a factor of about four during daytime. Recently one search coil magnetometer is being installed at Geomagnetic Research Laboratory Allahabad. This system will help in studying pulsations in various frequency bands in the low latitudes.

Topics of Investigations are Mainly

- The process of understanding the manifestation and development of the geomagnetic storm process under varied interplanetary conditions following rapid eruptions from the active Sun as solar flares and gigantic Coronal mass ejections (CMEs) etc.

- The effect of the impingement of the solar energetic particles (SEP), following active solar flare occurrences in the Sun and fast coronal mass ejections leading to dynamic changes in the earth's magnetosphere.
- Estimation of storm time energy budget by computing solar wind energies, magnetospheric coupling energies, auroral and Joule heating energies and also ring current energies.
- The extreme magnetic storm of September 1-2, 1859 was one of the most intense events in recorded history, based on the data deduced from the reported ground magnetic observations from the Colaba magnetic observatory. Using empirical results on the interplanetary magnetic field strengths of magnetic clouds versus velocity, it was shown that the well known September 1, 1859 Carrington solar flare most likely had an associated intense magnetic cloud ejection which led to the magnetic storm on Earth, which was observed as a large decrease in the Horizontal Component (Dst˜-1600 nT) during the main phase, as recorded at Colaba (Bombay).

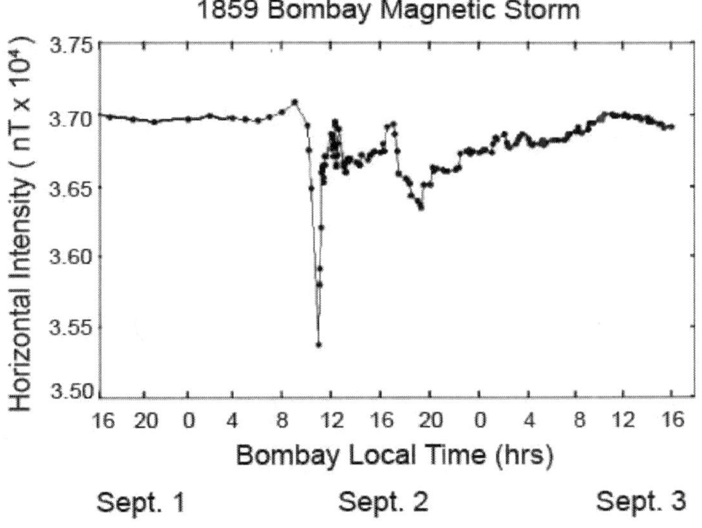

- The intense storm events on October 29 and October 31, 2003 resulted from earth directed strong eruption of Coronal Mass Ejections (CMEs) at a speed around 2000km/s. Due to the occurrence of large proton events following the flare events, solar wind detectors onboard NASA/ESA's ACE/SOHO satellites at L1 – Lagrangian point- were affected, thus reporting saturated values in the data stretch for sometime.

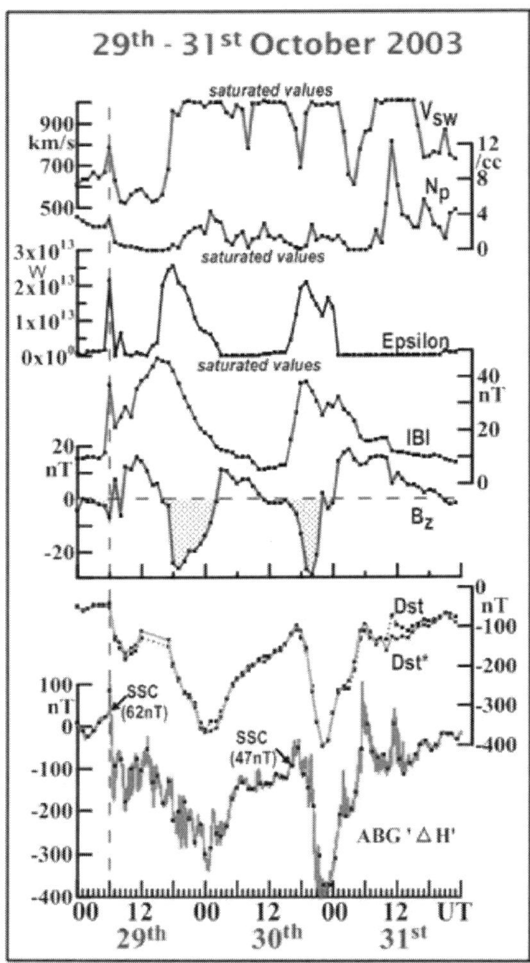

Figure: *displays the respective growth of the main phase events during the two storm sudden commencements occurred on October 29 and October 30, 2003, following the highly eruptive solar events. Solar wind and interplanetary magnetic field parameters (ACE) during the storm events are also given.*

- o During solar maximum, solar flares, geo-effective CMEs and the intense southward IMF are responsible for the development of non-recurrent geomagnetic storms. Dominance of high speed streams from coronal holes is a known feature during solar minimum. If the high speed streams overtake slower speed streams, the magnetic field and plasma compressions results at their interfaces known as Co-rotating Interaction Regions (CIRs) which are responsible for recurrent geomagnetic storms during the descending phase of the solar cycle. Multiple peak signatures of the occurrence of geomagnetic activity seem to dominate during recent solar cycles. Main sources of origin for these structured activity maxima are investigated by studying the variable characteristics observed in the low latitude storm development pattern.

- o Low latitude magnetograms often display sporadic positive increase in the Horizontal component of the geomagnetic field, named as positive bays and at times as negative bays. These bay structures are believed to be associated with the onset of the substorms and a quantitative estimate of the asymmetric development of the magnitude and phase of the bay events are brought out depending on the level of magnetic activity.

- Long series of magnetic data recorded at Indian magnetic observatories are used to quantify the occurrence characteristics of geomagnetic storms.

Atmosphere and Weather Phenomena

Weather is the state of the atmosphere, to the degree that it is hot or cold, wet or dry, calm or stormy, clear or cloudy. Weather, seen from an anthropological perspective, is something all humans in the world constantly experience through their senses, at least while being outside. There are socially and scientifically constructed understandings of what weather is, what makes it change, what effects it has on humans in different situations etc. Therefore weather is something people often communicate about. Turning back to the meteorological perspective, most weather phenomena occur in the troposphere, just below the stratosphere. Weather generally refers to day-to-day temperature and precipitation activity, whereas climate is the term for the statistics of atmospheric conditions over longer periods of time. When used without qualification, "weather", is generally understood to mean the weather of Earth.

Weather is driven by air pressure (temperature and moisture) differences between one place and another. These pressure and temperature differences can occur due to the sun angle at any particular spot, which varies by latitude from the tropics. The strong temperature contrast between polar and tropical air gives rise to the jet stream. Weather systems in the mid-latitudes, such as extratropical cyclones, are caused by instabilities of the jet stream flow. Because the Earth's axis is tilted relative to its orbital plane, sunlight is incident at different angles at different times of the year. On Earth's surface, temperatures usually range ±40 °C (–40 °F to 100 °F) annually. Over thousands of years, changes in Earth's orbit can affect the amount and distribution of solar energy received by the Earth, thus influencing long-term climate and global climate change.

Surface temperature differences in turn cause pressure differences. Higher altitudes are cooler than lower altitudes due to differences in compressional heating. Weather forecasting is the application of science and technology to predict the state of the atmosphere for a future time and a given location.

The system is a chaotic system; so small changes to one part of the system can grow to have large effects on the system as a whole. Human attempts to control the weather have occurred throughout human history, and there is evidence that human activities such as agriculture and industry have modified weather patterns. Studying how the weather works on other planets has been helpful in understanding how weather works on Earth. A famous landmark in the Solar System, Jupiter's *Great Red Spot*, is an anticyclonic storm known to have existed for at least 300 years. However, weather is not limited to planetary bodies. A star's corona is constantly being lost to space, creating what is essentially a very thin atmosphere throughout the Solar System. The movement of mass ejected from the Sun is known as the solar wind.

Causes

On Earth, the common weather phenomena include wind, cloud, rain, snow, fog and dust storms. Less common events include natural disasters such as tornadoes, hurricanes, typhoons and ice storms. Almost all familiar weather phenomena occur in the troposphere (the lower part of the atmosphere). Weather does occur in the stratosphere and can affect weather lower down in the troposphere, but the exact mechanisms are poorly understood.

Weather occurs primarily due to air pressure (temperature and moisture) differences between one place to another. These differences can occur due to the sun angle at any particular spot, which varies by latitude from the tropics. In other words, the farther from the tropics one lies, the lower the sun angle is, which causes those locations to be cooler due to the indirect sunlight. The strong temperature contrast between polar and tropical air gives rise to the jet stream. Weather systems in the mid-latitudes, such as extratropical cyclones, are caused by instabilities of the jet stream flow. Weather systems in the tropics, such as monsoons or organised thunderstorm systems, are caused by different processes.

Because the Earth's axis is tilted relative to its orbital plane, sunlight is incident at different angles at different times of the year. In June the Northern Hemisphere is tilted towards the sun, so at any given Northern Hemisphere latitude sunlight falls more directly on that spot than in December. This effect causes seasons. Over thousands to hundreds of thousands of years, changes in Earth's orbital parameters affect the amount and distribution of solar energy received by the Earth and influence long-term climate.

The uneven solar heating (the formation of zones of temperature and moisture gradients, or frontogenesis) can also be due to the weather itself in the form of cloudiness and precipitation. Higher altitudes are cooler than lower altitudes, which is explained by the lapse rate. On local scales, temperature differences can occur because different surfaces (such as oceans, forests, ice sheets, or man-made objects) have differing physical characteristics such as reflectivity, roughness, or moisture content.

Surface temperature differences in turn cause pressure differences. A hot surface heats the air above it raising the pressure of the air (Combined gas law) and causing it to expand which then lowers the air pressure and its density. The resulting horizontal pressure gradient accelerates the air from high to low pressure, creating wind, and Earth's rotation then causes curvature of the flow via the Coriolis effect. The simple systems thus formed can then display emergent behaviour to produce more complex systems and thus other weather phenomena. Large scale examples include the Hadley cell while a smaller scale example would be coastal breezes.

The atmosphere is a chaotic system, so small changes to one part of the system can grow to have large effects on the system as a whole. This makes it difficult to accurately predict weather more than a few days in advance, though weather forecasters are continually working to extend this limit through the scientific study of weather, meteorology. It is theoretically impossible to make useful day-to-day predictions more than about two weeks ahead, imposing an upper limit to potential for improved prediction skill.

Shaping the Planet Earth

Weather is one of the fundamental processes that shape the Earth. The process of weathering breaks down the rocks and soils into smaller fragments and then into their constituent substances. During rains precipitation, the water droplets absorb and dissolve carbon dioxide from the surrounding air. This causes the rainwater to be slightly acidic, which aids the erosive properties of water. The released sediment and chemicals are then free to take part in chemical reactions that can affect the surface further (such as acid rain), and sodium and chloride ions (salt) deposited in the seas/oceans. The sediment may reform in time and by geological forces into other rocks and soils. In this way, weather plays a major role in erosion of the surface.

Major Wind and Pressure Systems and Related Weather

Region	*Name*	*Pressure*	*Surface Winds*	*Weather*
Equator (0°)	Doldrums (ITCZ) (equatorial low)	Low	Light, variable winds	Cloudiness, abundant precipitation in all seasons; breeding ground for hurricanes. Relatively low sea-surface salinity because of high rainfall relative to evaporation
0°-30°N and S	Trade winds (easterlies)	-	Northeast in Northern Hemisphere; Southeast in Southern Hemisphere	Summer wet, winter dry; pathway for tropical disturbances
30°N and S	Horse latitudes	High	Light, variable winds	Little cloudiness; dry in all seasons. Relatively high sea-surface salinity because of high evaporation relative to precipitation
30°-60°N and S	Prevailing Westerlies	-	Southwest in Northern Hemisphere; Northwest in Southern Hemisphere	Winter wet, summer dry; pathway for subtropical high and low pressure
60°N and S	Polar front	Low	Variable	Stormy, cloudy weather zone; ample precipitation in all seasons
60°-90°N and S	Polar easterlies	-	Northeast in Northern Hemisphere; Southeast in Southern Hemisphere	Cold polar air with very low temperatures
90°N and S	Poles	High	Southerly in Northern Hemisphere; Northerly in Southern Hemisphere	Cold, dry air; sparse precipitation in all seasons

Effects on Populations

Weather has played a large and sometimes direct part in human history. Aside from climatic changes that have caused the gradual drift of populations (for example the desertification of the Middle East, and the formation of land bridges during glacial periods), extreme weather events have caused smaller scale population movements and intruded directly in historical events. One such event is the saving of Japan from invasion by the Mongol fleet of Kublai Khan by the Kamikaze winds in 1281. French claims to Florida came to an end in 1565 when a hurricane destroyed the French fleet, allowing Spain to conquer Fort Caroline. More recently, Hurricane Katrina redistributed over one million people from the central Gulf coast elsewhere across the United States, becoming the largest diaspora in the history of the United States.

The Little Ice Age caused crop failures and famines in Europe. The 1690s saw the worst famine in France since the Middle Ages. Finland suffered a severe famine in 1696–1697, during which about one-third of the Finnish population died.

Weather Forecasting

Weather forecasting is the application of science and technology to predict the state of the atmosphere for a given location. Human beings have attempted to predict the weather informally for millennia, and formally since the nineteenth century. Weather forecasts are made by collecting quantitative data about the current state of the atmosphere on a given place and using scientific understanding of atmospheric processes to project how the atmosphere will change.

Once an all-human endeavour based mainly upon changes in barometric pressure, current weather conditions, and sky condition, weather forecasting now relies on computer-based models that take many atmospheric factors into account. Human input is still required to pick the best possible forecast model to base the forecast upon, which involves pattern recognition skills, teleconnections, knowledge of model performance, and knowledge of model biases. The chaotic nature of the atmosphere, the massive computational power required to solve the equations that describe the atmosphere, error involved in measuring the initial conditions, and an incomplete understanding of atmospheric processes mean that forecasts become less accurate as the difference in current time and the time for which the forecast is being made (the *range* of the forecast) increases. The use of ensembles and model consensus help narrow the error and pick the most likely outcome.

There are a variety of end uses to weather forecasts. Weather warnings are important forecasts because they are used to protect life and property. Forecasts based on temperature and precipitation are important to agriculture, and therefore to traders within commodity markets. Temperature forecasts are used by utility companies to estimate demand over coming days. On an everyday basis, people use weather forecasts to determine what to wear on a given day. Since outdoor activities are severely curtailed by heavy rain, snow and the wind chill, forecasts can be used to plan activities around these events, and to plan ahead and survive them.

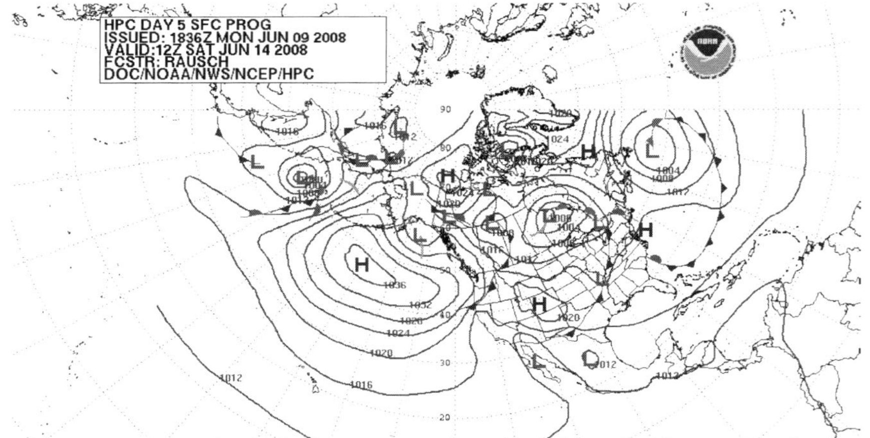

Figure: *Forecast of surface pressures five days into the future for the north Pacific, North America, and north Atlantic ocean.*

It was not until the invention of the electric telegraph in 1835 that the modern age of weather forecasting began. Before that, the fastest that distant weather reports could travel was around 100 miles per day (160 km/d), but was more typically 40–75 miles per day (60–120 km/day) (whether by land or by sea). By the late 1840s, the telegraph allowed reports of weather conditions from a wide area to be received almost instantaneously, allowing forecasts to be made from knowledge of weather conditions further upwind.

The two men credited with the birth of forecasting as a science were officer of the Royal Navy Francis Beaufort and his protégé Robert FitzRoy. Both were influential men in British naval and governmental circles, and though ridiculed in the press at the time, their work gained scientific credence, was accepted by the Royal Navy, and formed the basis for all of today's weather forecasting knowledge.

Beaufort developed the Wind Force Scale and Weather Notation coding, which he was to use in his journals for the remainder of his life. He also promoted the development of reliable tide tables around British shores, and with his friend William Whewell, expanded weather record-keeping at 200 British Coast guard stations.

Robert FitzRoy was appointed in 1854 as chief of a new department within the Board of Trade to deal with the collection of weather data at sea as a service to mariners. This was the forerunner of the modern Meteorological Office. All ship captains were tasked with collating data on the weather and computing it, with the use of tested instruments that were loaned for this purpose.

A terrible storm in 1859 that caused the loss of the *Royal Charter* inspired FitzRoy to develop charts to allow predictions to be made, which he called *"forecasting the weather"*, thus coining the term "weather forecast". Fifteen land stations were established to use the new telegraph to transmit to him daily reports of weather at set times leading to the first gale warning service. His warning service for shipping was initiated in February 1861, with the use of telegraph communications. In the following year a system was introduced of hoisting storm warning cones at the principal ports when a gale was expected. The *"Weather Book"* which FitzRoy published in 1863 was far in advance of the scientific opinion of the time.

The electric telegraph network became denser in the 1870s, allowing for the more rapid dissemination of warnings; this also led to the development of an observational network which could then be used to provide synoptic analyses. To convey accurate information, it soon became necessary to have a standard vocabulary describing clouds; this was achieved by means of a series of classifications first achieved by Luke Howard in 1802, and standardized in the *International Cloud Atlas* of 1896.

Numerical Prediction

It was not until the 20th century that advances in the understanding of atmospheric physics led to the foundation of modern numerical weather prediction. In 1922, English scientist Lewis Fry Richardson published "Weather Prediction By Numerical Process", after finding notes and derivations he worked on as an ambulance driver in World War I. He described therein how small terms in the prognostic fluid dynamics equations governing atmospheric flow could be neglected, and a finite differencing scheme in time and space could be devised, to allow numerical prediction solutions to be found.

Richardson envisioned a large auditorium of thousands of people performing the calculations and passing them to others. However, the sheer number of calculations required was too large to be completed without the use of computers, and the size of the grid and time steps led to unrealistic results in deepening systems. It was later found, through numerical analysis, that this was due to numerical instability. The first computerised weather forecast was performed by a team led by the mathematician John von Neumann; von Neumann publishing the paper *Numerical Integration of the Barotropic Vorticity Equation* in 1950. Practical use of numerical weather prediction began in 1955, spurred by the development of programmable electronic computers.

Broadcasts

The first ever daily weather forecasts were published in *The Times* on 1 August 1861, and the first weather maps were produced later in the same year. In 1911, the Met Office began issuing the first marine weather forecasts via radio transmission. These included gale and storm warnings for areas around Great Britain. In the United States, the first public radio forecasts were made in 1925 by Edward B. "E.B." Rideout, on WEEI, the Edison Electric Illuminating station in Boston. Rideout came from the U.S. Weather Bureau, as did WBZ weather forecaster G. Harold Noyes in 1931.

The world's first televised weather forecasts, including the use of weather maps, were experimentally broadcast by the BBC in 1936. This was brought into practice in 1949 after World War II. George Cowling gave the first weather forecast while being televised in front of the map in 1954. In America, experimental television forecasts were made by James C Fidler in Cincinnati in either 1940 or 1947 on the DuMont Television Network. In the late 1970s and early 80s, John Coleman, the first weatherman on ABC-TV's Good Morning America, pioneered the use of on-screen weather satellite information and computer graphics for television forecasts. Coleman was a co-founder of The Weather Channel (TWC) in 1982. TWC is now a 24-hour cable network.

Specialist Forecasting

There are a number of sectors with their own specific needs for weather forecasts and specialist services are provided to these users.

Air Traffic

Because the aviation industry is especially sensitive to the weather, accurate weather forecasting is essential. Fog or exceptionally low ceilings can prevent many aircraft from landing and taking off. Turbulence and icing are also significant in-flight hazards. Thunderstorms are a problem for all aircraft because of severe turbulence due to their updrafts and outflow boundaries, icing due to the heavy precipitation, as well as large hail, strong winds, and lightning, all of which can cause severe damage to an aircraft in flight. Volcanic ash is also a significant problem for aviation, as aircraft can lose engine power within ash clouds. On a day-to-day basis airliners are routed to take advantage of the jet stream tailwind to improve fuel efficiency. Aircrews are briefed prior to takeoff on the conditions to expect en route and at their destination. Additionally, airports often change which runway is being used to take advantage of a headwind. This reduces the distance required for takeoff, and eliminates potential crosswinds.

Marine

Commercial and recreational use of waterways can be limited significantly by wind direction and speed, wave periodicity and heights, tides, and precipitation. These factors can each influence the safety of marine transit. Consequently, a variety of codes have been established to efficiently transmit detailed marine weather forecasts to vessel pilots via radio, for example the MAFOR (marine forecast). Typical weather forecasts can be received at sea through the use of RTTY, Navtex and Radiofax.

Agriculture

Farmers rely on weather forecasts to decide what work to do on any particular day. For example, drying hay is only feasible in dry weather. Prolonged periods of dryness can ruin cotton, wheat, and corn crops. While corn crops can be ruined by drought, their dried remains can be used as a cattle feed substitute in the form of silage. Frosts and freezes play havoc with crops both during the spring and fall. For example, peach trees in full bloom can have their potential peach crop decimated by a spring freeze. Orange groves can suffer significant damage during frosts and freezes, regardless of their timing.

Forestry

Weather forecasting of wind, precipitations and humidity is essential for preventing and controlling wildfires. Different indices, like the *Forest fire weather index* and the *Haines Index*, have been developed to predict the areas more at risk to experience fire from natural or human causes. Conditions for the development of harmful insects can be predicted by forecasting the evolution of weather, too.

Utility Companies

Electricity and gas companies rely on weather forecasts to anticipate demand which can be strongly affected by the weather. They use the quantity termed the degree day to determine how strong of a use there will be for heating (heating degree day) or cooling (cooling degree day). These quantities are based on a daily average temperature of 65 °F (18 °C). Cooler temperatures force heating degree days (one per degree Fahrenheit), while warmer temperatures force cooling degree days. In winter, severe cold weather can cause a surge in demand as people turn up their heating. Similarly, in summer a surge in demand can be linked with the increased use of air conditioning systems in hot weather. By anticipating a surge in demand, utility companies can purchase additional supplies of power or natural gas before the price increases, or in some circumstances, supplies are restricted through the use of brownouts and blackouts.

Other Commercial Companies

Increasingly, private companies pay for weather forecasts tailored to their needs so that they can increase their profits or avoid large losses. For example, supermarket chains may change the stocks on their shelves in anticipation of different consumer spending habits in different weather conditions. Weather forecasts can be used to invest in the commodity market, such as futures in oranges, corn, soybeans, and oil.

Weather Modification

The aspiration to control the weather is evident throughout human history: from ancient rituals intended to bring rain for crops to the U.S. Military Operation Popeye, an attempt to disrupt supply lines by lengthening the North Vietnamese monsoon. The most successful attempts at influencing weather involve cloud seeding; they include the fog- and low stratus dispersion techniques employed by major airports, techniques used to increase winter precipitation over mountains, and techniques to suppress hail. A recent example of weather control was China's preparation for the 2008 Summer Olympic Games. China shot 1,104 rain dispersal rockets from 21 sites in the city of Beijing in an effort to keep rain away from the opening ceremony of the games on 8 August 2008. Guo Hu, head of the Beijing Municipal Meteorological Bureau (BMB), confirmed the success of the operation with 100 millimeters falling in Baoding City of Hebei Province, to the southwest and Beijing's Fangshan District recording a rainfall of 25 millimeters.

Whereas there is inconclusive evidence for these techniques' efficacy, there is extensive evidence that human activity such as agriculture and industry results in inadvertent weather modification:

- Acid rain, caused by industrial emission of sulfur dioxide and nitrogen oxides into the atmosphere, adversely affects freshwater lakes, vegetation, and structures.

- Anthropogenic pollutants reduce air quality and visibility.
- Climate change caused by human activities that emit greenhouse gases into the air is expected to affect the frequency of extreme weather events such as drought, extreme temperatures, flooding, high winds, and severe storms. However, some experts argue these claims are unfounded and take issue with these conclusions.
- Heat, generated by large metropolitan areas have been shown to minutely affect nearby weather, even at distances as far as 1,600 kilometres (990 mi).

The effects of inadvertent weather modification may pose serious threats to many aspects of civilization, including ecosystems, natural resources, food and fiber production, economic development, and human health.

Microscale Meteorology

Microscale meteorology is the study of short-lived atmospheric phenomena smaller than mesoscale, about 1 km or less. These two branches of meteorology are sometimes grouped together as "mesoscale and microscale meteorology" (MMM) and together study all phenomena smaller than synoptic scale; that is they study features generally too small to be depicted on a weather map. These include small and generally fleeting cloud "puffs" and other small cloud features.

Extremes on Earth

On Earth, temperatures usually range ±40 °C (100 °F to –40 °F) annually. The range of climates and latitudes across the planet can offer extremes of temperature outside this range. The coldest air temperature ever recorded on Earth is –89.2 °C (–128.6 °F), at Vostok Station, Antarctica on 21 July 1983. The hottest air temperature ever recorded was 57.7 °C (135.9 °F) at 'Aziziya, Libya, on 13 September 1922, but that reading is queried. The highest recorded average annual temperature was 34.4 °C (93.9 °F) at Dallol, Ethiopia. The coldest recorded average annual temperature was –55.1 °C (–67.2 °F) at Vostok Station, Antarctica.

The coldest average annual temperature in a permanently inhabited location is at Eureka, Nunavut, in Canada, where the annual average temperature is –19.7 °C (–3.5 °F).

Mawsynram : The Wettest Place on Earth

Mawsynram is a village in the East Khasi Hills district of Meghalaya state in north-eastern India, 65 kilometres from Shillong.

It is reportedly the wettest place on Earth, with an annual rainfall of 11,872 millimetres (467.4 in), but that claim is disputed by Llorz, Colombia, which had an average yearly rainfall of 12,717 millimetres (500.7 in) between 1952 and 1989 and Lopez del Micay, also in Colombia, with 12892 mm between 1960 and 2012. According to the *Guinness Book of World Records*, Mawsynram received 26,000 millimetres (1,000 in) of rainfall in 1985.

Mawsynram is located at 25° 18' N, 91° 35' E, at an altitude of about 1,400 metres (4,600 ft), 16 km west of Cherrapunji, in the Khasi Hills in the state of Meghalaya (India) . The name of the village contains *Maw*, a Khasi word meaning *stone*, and thus might refer to certain megaliths in the surrounding area. Khasi Hills are rich with such megaliths.

Climate and Rainfall

Under the Köppen climate classification, Mawsynram features a subtropical highland climate with an extraordinarily rainy and lengthy monsoonal season. Based on the data of a recent few decades, it appears to be the wettest place in the world, or the place with the highest average annual rainfall. Mawsynram receives nearly 12 m of rain in an average year, and the vast majority of it falls during the monsoon months.

Primarily due to the high altitude, it seldom gets truly hot in Mawsynram. Average monthly temperatures range from around 10 degrees Celsius in January to just above 20 degrees Celsius in August. The village also experiences a brief but noticeably drier season from December until February, when monthly precipitation on average does not exceed 60 mm. The relative dearth of precipitation during the village's "low sun" season is a trait shared by many areas with this climate.

Coldest Places

Vostok Station, Antarctica: Vostok Station in Antarctica is one of the coldest places on Earth. The Russian research station is where the lowest recorded temperature was taken at -89.2 degrees Celsius (-128.56 Fahrenheit). Antarctica is not only the coldest place but also the windiest and has the highest overall elevation of any continent in the world. Despite the fact that Antarctica contains 90 percent of the ice in the world, it receives almost no rainfall, making it technically a desert.

Oymyakon, Russia: Everyone knows Russia is cold. Oymyakon, Russia is the coldest inhabited place in the world. The lowest temperature recorded there is -71.2 degrees Celsius (-96.16 Fahrenheit).

This small town has a population of approximately 500 people. Although it's very cold there in winter, people still continue to live their lives and students attend school, unless it gets lower than -52 degrees Celsius (-61.6 Fahrenheit). Sometimes it gets so cold at night that birds freeze.

Verkhoyansk, Russia: Another very cold town in Russia, Verkhoyansk, has a population of around 1,400 inhabitants. The lowest temperature ever recorded there is -69.8 degrees Celsius (-93.64 Fahrenheit). Between the 1860s and early 20th century Verkhoyansk was used to house political exiles.

Hottest Places

Death Valley, California: The name of the place says it all. Located in Southeastern California, Death Valley is the hottest, driest, and lowest place in North America. The hottest temperature recorded in Death Valley is 56.7 degrees Celsius (134.06 Fahrenheit). Although Death Valley is very hot during the day, at night the temperature often drops below freezing.

Al'Aziziyah, Libya: Al'Aziziyah, located in Northwest Libya, is one of the hottest habitable places on Earth. On Sept. 13, 1922, a high temperature of 57.8 degrees Celsius (136.04 Fahrenheit) was recorded in Al'Aziziyah. Later, experts discounted the reading, mainly because they said the person who took it was untrained, but the area is still considered one of the hottest in the world.

Ghudamis, Libya: Libya is a hot place; Ghudamis is an oasis town on the west of Libya. The highest temperature recorded in Ghudamis is 55 degrees Celsius (131 Fahrenheit). People build their houses out of lime, palm tree trunks, and mud in order to survive the brutal heat and sandstorms.

Extraterrestrial within the Solar System

Studying how the weather works on other planets has been seen as helpful in understanding how it works on Earth. Weather on other planets follows many of the same physical principles as weather on Earth, but occurs on different scales and in atmospheres having different chemical composition. The Cassini–Huygens mission to Titan discovered clouds formed from methane or ethane which deposit rain composed of liquid methane and other organic compounds. Earth's atmosphere includes six latitudinal circulation zones, three in each hemisphere. In contrast, Jupiter's banded appearance shows many such zones, Titan has a single jet stream near the 50th parallel north latitude, and Venus has a single jet near the equator.

One of the most famous landmarks in the Solar System, Jupiter's *Great Red Spot*, is an anticyclonic storm known to have existed for at least 300 years. On other gas giants, the lack of a surface allows the wind to reach enormous speeds: gusts of up to 600 metres per second (about 2,100 km/h or 1,300 mph) have been measured on the planet Neptune. This has created a puzzle for planetary scientists. The weather is ultimately created by solar energy and the amount of energy received by Neptune is only about $^{1}D_{900}$ of that received by Earth, yet the intensity of weather phenomena on Neptune is far greater than on Earth. The strongest planetary winds discovered so far are on the extrasolar planet HD 189733 b, which is thought to have easterly winds moving at more than 9,600 kilometres per hour (6,000 mph).

Space Weather

Space weather is a branch of space physics and aeronomy concerned with the time varying conditions within the solar system, including the solar wind, and especially the space surrounding the Earth, including conditions in the magnetosphere, ionosphere, and thermosphere. Space weather is distinct from the terrestrial weather of the Earth's atmosphere (troposphere and stratosphere). The science of space weather is focused in two distinct directions: fundamental research and practical applications. The term *space weather* was not used until the 1990s.

For many centuries, the effects of space weather were noticed but not understood. Beautiful displays of auroral light have long been admired by people living at high latitudes. In 1724, George Graham reported that the needle of a magnetic compass was regularly deflected from magnetic north over the course of each day. This effect was eventually attributed to overhead electric currents flowing in what we now recognise as the ionosphere and magnetosphere by Balfour Stewart in 1882 and, subsequently, confirmed by Arthur Schuster in 1889 from analysis of magnetic observatory data. In 1852, the astronomer and British major general, Edward Sabine showed that the probability of the occurrence of magnetic storms on Earth is correlated with the number of sunspots, thus demonstrating the existence of what we, today, might call solar-terrestrial interaction. In 1859, a great magnetic storm caused brilliant auroral displays, and it also disrupted telegraph operations around the world, events that were reported in many major newspapers at that time. Richard Carrington correctly connected the storm with a solar flare that he had observed the day before in the vicinity of a large sunspot group — thus demonstrating that specific events occurring on the Sun's surface can, in turn, affect the Earth.

Kristian Birkeland explained the physics of aurora by creating artificial aurora in his laboratory and predicted the solar wind. With the introduction of radio for commercial and military uses, it was noted that periods of extreme static or noise occurred. Severe radar jamming during a large solar event in 1942 led to the discovery of solar radio bursts (radio waves which cover a broad frequency range created by a solar flare), another aspect of space weather.

In the 20th century, the interest in space weather has expanded as military and commercial systems have come to depend on systems affected by space weather. Communications satellites are a vital part of global commerce. Weather satellite systems provide information about terrestrial weather. The signals from satellites of the Global Positioning System are used in a wide variety of commercial products and processes. Space weather phenomena can interfere with or damage these satellites or interfere with the radio signals to and from these satellites. Space weather phenomena can cause damaging surges in long electrical transmission lines and expose passengers and crew of aircraft travel to radiation, especially on polar routes.

The International Geophysical Year (IGY), created an enormous increase in research into space weather. Ground-based data obtained during IGY demonstrated that the aurora occurred in an *auroral oval*, a permanent region of luminescence 15 to 25 degrees in latitude from the magnetic poles and 5 to 20 degrees wide. In 1958, the Explorer I satellite discovered the Van Allen belts or regions of radiation particles trapped by the Earth's magnetic field. In January 1959, the Soviet satellite Luna 1 first directly observed the solar wind and measured its strength. In 1969, INJUN-5 (a.k.a. Explorer 40) made the first direct observation of the electric field impressed on the Earth's high latitude ionosphere by the solar wind. In the early 1970's, Triad data demonstrated that permanent electric currents flowed between the auroral oval and the magnetosphere. From these and other fundamental discoveries, research into space weather has grown exponentially.

Within our own solar system, space weather is greatly influenced by the speed and density of the solar wind and the interplanetary magnetic field (IMF) carried by the solar wind plasma. A variety of physical phenomena are associated with space weather, including geomagnetic storms and substorms, energization of the Van Allen radiation belts, ionospheric disturbances and scintillation of satellite-to-ground radio signals and long-range radar signals, aurora and geomagnetically induced currents at Earth's surface. Coronal mass ejections and their associated shock waves are also important drivers of space weather as they can compress the magnetosphere and trigger geomagnetic storms. Solar energetic particles, accelerated by coronal mass ejections or solar flares, are also an important driver of space weather as they can damage electronics onboard spacecraft (e.g. Galaxy 15 failure), and threaten the life of astronauts.

The term space weather came into usage in the 1990s when it became apparent that the impact of the space environment on human systems demanded a more coordinated research and application framework. The purpose of the National Space Weather Program in the USA is to focus research on the needs of the commercial and military communities which are affected by space weather, to connect the research community to the user community, to create coordination between operational data centers and to create better definitions of what the user community needs are. The concept was turned into an action plan in 2000, an implementation plan in 2002, an assessment in 2006 and a revised strategic plan in 2010. A revised action plan will be released in 2011 and a revised implementation plan will be release in 2012. One part of the National Space Weather Program is to make users aware that space weather affects their business. Private companies now acknowledge space weather "is a real risk for today's businesses".

Effect of Space Weather on Space Systems

Spacecraft Anomalies

Spacecraft malfunction for a variety of reasons. Some malfunctions are reported but many are not reported. A few failures can be directly attributed to space weather; many more failures are suspected to have a space weather component; and many failures are unrelated to space weather. One indicator that space weather is a significant driver of spacecraft failure is that 46 of the 70 failures reported in 2003 occurred during the October 2003 geomagnetic storm. The two most common adverse space weather effects on spacecraft are radiation damage and spacecraft charging. Radiation (high energy particles) passes through the skin of the spacecraft and into the electronic components. In most cases the radiation causes an erroneous signal or changes one bit in memory of a spacecraft's electronics (single event upsets). In a few cases, the radiation destroys a section of the electronics (single-event latchup). Spacecraft charging is the accumulation of an electrostatic charge on a non-conducting material on the spacecraft's surface by low energy particles. If enough charge is built-up, a discharge (spark) occurs.

Damage to the spacecraft is done by causing an erroneous signal to be detected and acted on by the spacecraft computer as if the signal came from the ground controller or the electronics are damaged by a surge of electrical current. A recent study indicates that spacecraft charging is the predominant space weather effect on spacecraft in geosynchronous orbit.

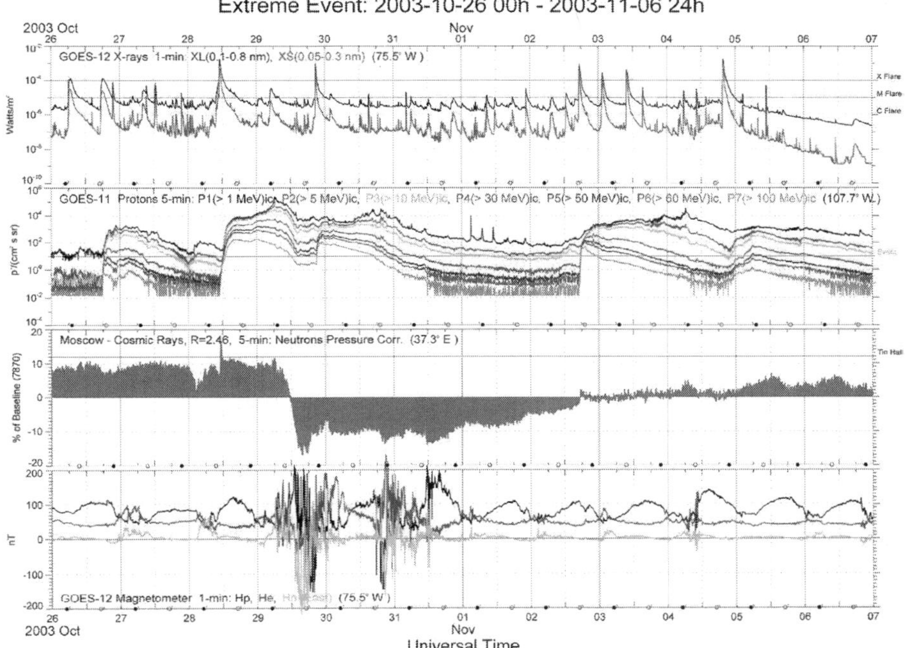

***Figure:** GOES-11 and GOES-12 monitored space weather conditions during the October 2003 solar activity.*

Spacecraft Orbit Changes

The orbits of spacecraft in low Earth orbit (LEO) decay to lower and lower altitudes due to the resistance from the friction between the spacecraft's surface (*i.e.* , drag) and the outer layer of the Earth's atmosphere (a.k.a. the thermosphere and exosphere). Eventually, a spacecraft's orbit will decay so much that it will fall out of orbit and crash to the Earth's surface. Many spacecraft launched in the past couple of decades have the ability to fire a small rocket:

(1) to increase the altitude to compensate for the decay and extend the lifetime in space,

(2) to re-enter the atmosphere and crash into the ocean, or

(3) change the orbit to avoid collision with other spacecraft.

In order to accomplish the goal of firing a small rocket, very precise information about the orbit is needed. A geomagnetic storm can cause an orbit change over a couple of days that otherwise would occur over a year or more. The geomagnetic storm adds heat to the thermosphere, causing the thermosphere to expand and rise, which increases the drag on spacecraft in low Earth orbits. The 2009 satellite collision between the Iridium 33 and Cosmos 2251 demonstrated the importance of having precise knowledge of all objects in orbit. Iridium 33 had the capability to maneuver out of the path of Cosmos 2251 and could have evaded the crash, if a credible collision prediction had been available,

Effect of Radiation on Humans in Space

The exposure of a human body to ionizing radiation has the same harmful effects whether the source of the radiation is a medical X-ray machine, a nuclear power plant or radiation in space. The degree of the harmful effect depends on the length of exposure and the energy density of the radiation. The ever-present radiation belts extend down to the altitude of manned spacecraft such as the International Space Station (ISS) and the Space Shuttle but the amount of exposure is within the acceptable lifetime exposure limit under normal conditions. During a major space weather event which includes a burst of solar energetic particles, the flux can increase by one to several orders of magnitude. There are areas within ISS where the thickness of the spacecraft surface and equipment can provide extra shielding and may keep the total dose absorbed within lifetime safe limits. For the Shuttle, such an event would have required an immediate termination of the mission.

Effects of Space Weather on Ground Systems

Disruption of GPS and Other Spacecraft Signals

The ionosphere bends radio waves in the same manner that water in a swimming pool bends visible light. When the medium through which the light or radio waves travel is disturbed, the light image or radio information is distorted and can become unrecognisable. The degree of distortion (scintillation) of a radio wave by the ionosphere depends on the frequency of the radio signal. Radio signals in the VHF band (30 to 300 MHz) can be distorted beyond recognition by a disturbed ionosphere. Radio signals in the UHF band (300 MHz to 3 GHz) will propagate through a disturbed ionosphere but a receiver may not be able to keep locked to the carrier frequency. The Global Positioning System uses signals at 1575.42 MHz (L1) and 1227.6 MHz (L2) which can be distorted by a disturbed ionosphere and a receiver computes an erroneous position or fails to compute any position. Because the GPS signals are used by wide range of applications, any space weather event which makes GPS signal unreliable, the impact on society can be significant. For example the Wide Area Augmentation System (WAAS) operated by the Federal Aviation Administration is used as a precision navigation tool for commercial aviation in North America. It is disabled by every major space weather event. In some cases WAAS is disabled for minutes and in a few cases it has been disabled for a few days. Major space weather events can push the disturbed polar ionosphere 10° to 30° of latitude toward the equator and can cause large ionospheric gradients (changes in density over distance of 100's of km) at mid and low latitude. Both of these factors can distort GPS signals.

Disruption of Long-Distance Radio Signals

Radio wave in the HF band (3 to 30 MHz) (also known as the shortwave band) are bent so much by the ionosphere that they are reflected back in the same manner as a mirror reflects light. Since the ground also reflects HF wave, a signal can be transmitted around the curvature of the Earth to a distant station. During the 20th century, HF communications was the only method for a ship or aircraft far from land or a base station to communicate. With the advent of systems such as Iridium, there are now other methods of communications but HF is still considered to be critical because not all vessels carry the newer equipment and even if the newer equipment is on board, HF is considered a critical backup system. Space weather events can create irregularities in the ionosphere that scatter HF signals instead of reflecting them and make HF communications over long distance poor or impossible. At auroral and polar latitudes, small space weather events which occur frequently disrupt HF communications. At mid-latitudes, HF communications are disrupted by solar radio bursts, by X-rays from solar flares (which enhance and disturb the ionospheric D-layer) and by TEC enhancements and irregularities during major geomagnetic storms which are infrequent.

Transpolar routes flown by airplanes are particularly sensitive to space weather, in part because of Federal Aviation Regulations requiring reliable communication over the entire flight. It is estimated to cost about $100,000 each time such a flight is diverted from a polar route.

Effect of Radiation on Humans at and Near Ground Level

The Earth's magnetic field guides cosmic ray and solar energetic particles to polar latitudes and radiation particles enter the mesosphere and stratosphere. Cosmic rays at the top of the atmosphere shatter atmospheric atoms and create lower energy, but still harmful, radiation particles which penetrate deep into the atmosphere. All aircraft flying above 10 km (33,000 feet) altitude are exposed to a noticeable amount of radiation. The exposure is greater in polar regions than at mid-latitude and equatorial regions. Many commercial aircraft from Europe and North America to East Asia fly over the polar region. When a space weather event causes radiation exposure to exceed the safe level set by aviation authorities, the aircraft's flight path is deviated to avoid the polar region.

Ground Induced Electric Fields

Magnetic storm activity can induce geoelectric fields in the Earth's conducting lithosphere. Corresponding voltage differentials can find their way into electric power grids through ground connections, driving uncontrolled electric currents that interfere with grid operation, damaging transformers, tripping protective relays, and sometimes causing blackouts. The reality of this complicated chain of causes and effects was demonstrated during the great magnetic storm of March 1989, which caused the complete collapse of the Hydro-Québec electric-power grid in Canada, temporarily leaving nine million people without electricity. The possible occurrence in the future of an even more intense magnetic storm, one that could cause widespread loss of electric power, has motivated regulatory agencies to issue operational standards intended to mitigate induction-hazard risks. Concerns in the private sector have motivated reinsurance companies to commission related assessments of risk.

Geophysical Exploration

Air and ship borne magnetic surveys can be affected by rapid magnetic field variations during geomagnetic storms. Geomagnetic storms cause data interpretation problems because the space-weather-related magnetic field changes are similar in magnitude to those of the sub-surface crustal magnetic field in the survey area. Accurate geomagnetic storm warnings, including an assessment of the magnitude and duration of the storm, allows for an economic use of survey equipment.

Geophysics and Hydrocarbon Production

For economic and other reasons, oil and gas production often involves the directional drilling of well paths many kilometers from a single wellhead in both the horizontal and vertical directions. Accuracy requirements are strict, due to target size – reservoirs may only be a few tens to hundreds of meters across – and for safety reasons, because of the proximity of other boreholes. Surveying by the most accurate gyroscopic method is expensive, since it can involve the cessation of drilling for a number of hours. An alternative is to use a magnetic survey, which enables measurement while drilling (MWD). Near real time magnetic data can be used to correct the drilling direction and nearby magnetic observatories prove vital. Magnetic data and space weather forecasts can also be helpful in clarifying unknown sources of drilling error on an on-going basis.

Effect of Space Weather on Terrestrial Weather

The amount of energy entering the troposphere and stratosphere from all space weather phenomena is trivial compared to the solar insolation in the visible and infrared portions of the solar electromagnetic spectrum. However there does seem to be some linkage between the 11-year sunspot cycle and the Earth's climate. For example, the Maunder minimum, a 70-year period almost devoid of sunspots, correlates to a cooling of the Earth's climate. One suggestion for the linkage between space and terrestrial weather is that changes in cosmic ray flux cause changes in the amount of cloud formation. Another suggestion is that variations in the EUV flux subtly influence existing drivers of the climate and tips the balance between states such as the El Niño/La Niña states. However, a linkage between space weather and the climate has not been demonstrated conclusively.

Observations of Space Weather

The observation of space weather is done both for scientific research and for applications. The type of observation done for science has varied over the years as the frontiers of our understanding has increased and due to competition for resources from other types of space-related research. The observations related to applications have been more systematic and has expanded over the years as awareness and applications have increased.

Observing Space Weather from the Ground

Presently, space weather is monitored at ground level by observing changes in the Earth's magnetic field over periods of seconds to days, by observing the surface of the Sun and by observing radio noise created in the Sun's atmosphere.

The Sunspot Number (SSN) is the number of sunspots on the Sun's photosphere in visible light on the side of the Sun visible to an Earth observer. The number and total area of sunspots are related to the brightness of the Sun in the extreme ultraviolet (EUV) and X-ray portions of the solar spectrum and to solar activity such as solar flares and coronal mass ejections (CMEs). 10.7 cm radio flux (F10.7) is a measurement of RF emissions from the Sun and is approximately correlated with the solar EUV flux. Since this RF emission is easily obtained from the ground and EUV flux is not, this value has been measured and disseminated continuously since 1947. The world standard measurements are made by the Dominion Radio Astrophysical Observatory at Penticton, B.C., Canada and reported once a day at local noon in solar flux units (10^{-22}W $\cdot$m^{-2} $\cdot$Hz^{-1}). F10.7 is archived by the National Geophysical Data Center.

Fundamental space weather monitoring data are provided by ground-based magnetometers and magnetic observatories. Indeed, magnetic storms were first discovered by ground-based measurement of occasional magnetic disturbance. Ground magnetometer data are used for informing real-time situational awareness, for post-event analysis of effects, and because many magnetic observatories have been in continuous operations for decades to centuries, their data also inform studies of long-term changes in space climatology.

Dst index is an estimate of the magnetic field change at the Earth's magnetic equator due to a ring of electrical current at and just earthward of GEO. The index is based on data from four ground-based magnetic observatories between 21° and 33° magnetic latitude during a one-hour period. Stations closer to the magnetic equator are not used due to ionospheric effects. The Dst index is compiled and archived by the World Data Center for Geomagnetism, Kyoto.

Kp/ap Index: 'a' is an index created from the geomagnetic disturbance at one mid-latitude (40° to 50° latitude) geomagnetic observatory during a 3-hour period. 'K' is the quasi-logarithmic counterpart of the 'a' index. Kp and ap are the average of K and an over 13 geomagnetic observatories to represent planetary-wide geomagnetic disturbances. The Kp/ap index indicates both geomagnetic storms and substorms (auroral disturbance). Kp/ap is available from 1932 onward.

AE index is compiled from geomagnetic disturbances at 12 geomagnetic observatories in and near the auroral zones and is recorded at 1-minute intervals. The AE index is made public with a delay of two to three days, which severely limits its utility for space weather applications. The AE index indicates the intensity of geomagnetic substorms except during a major geomagnetic storm when the auroral zones expand equatorward from the observatories.

Radio noise burst are observed and reported by the Radio Solar Telescope Network to the U.S. Air Force and to NOAA. The radio bursts are associated with plasma from a solar flare interacting with the ambient solar atmosphere.

The Sun's photosphere is observed continuously by a series of observatories for activity which can be the precursors to solar flares and CMEs. The Global Oscillation Network Group (GONG) project monitors both the surface and the interior of the Sun by using helioseismology, the study of sound waves propagating through the Sun and observed as ripples on the solar surface. GONG can detect sunspot groups on the far side of the Sun. This ability has recently been verified by visual observations from the NASA STEREO spacecraft.

Neutron monitors on the ground indirectly monitor cosmic rays from the Sun and galactic sources. Cosmic rays do not reach the Earth's surface due to the shielding of the Earth's magnetic field and atmosphere. When cosmic rays interact with the atmosphere, atomic interactions occur which cause a shower of lower energy particles to descend deeper into the atmosphere and to ground level. The presence of cosmic rays in the near-Earth space environment can be detected by monitoring high energy neutrons at ground level. Small fluxes of cosmic rays are present continuously. Large fluxes are produced by the Sun during events related to energetic solar flares.

Total Electron Content (TEC) is a measure of the ionosphere over a given location. TEC is the number of electrons in a column one meter square from the base of the ionosphere (approximately 90 km altitude) to the top of the ionosphere (approximately 1000 km altitude). Many of the measurements of TEC are made by monitoring the two frequencies transmitted by GPS spacecraft. Presently GPS TEC is monitored and distributed in real time from more than 360 stations maintained by numerous agencies in many countries.

Geoeffectiveness is a measure of how strongly the magnetic fields of space weather events, such as coronal mass ejections, will couple with the Earth's magnetic field. This is determined by the direction the magnetic field held within the plasma that originates from the sun. New techniques measuring Faraday Rotation in radio waves are being developed to measure the direction of the magnetic field.

Observing Space Weather with Satellites

After Explorer I discovered that space was not a void, many research spacecraft have been launched to discover and characterize the space environment. There have been too many spacecraft since then to list them all here and they have carried a wide variety of instruments. The spacecraft of the Orbiting Geophysical Observatory series were among the first spacecraft with the mission of discovering the space environment. Significant recent spacecraft are the NASA-ESA Solar-Terrestrial Relations Observatory (STEREO) pair of spacecraft launched in 2006 into solar orbit and the Van Allen Probes, launched in 2012 into a highly elliptical Earth-orbit. The two STEREO spacecraft drift away from the earth by about 50° per year, one leading and the other trailing the earth in its orbit. Together they compile information about the Sun's surface and atmosphere in three dimensions. The Van Allen probes are obtaining detailed information about the radiation belts, geomagnetic storms and the relationship between the two.

The mission of most spacecraft is unrelated to gathering information about the space environment for research or applications, but some of these other spacecraft have carried auxiliary instrument or had some part of their primary payload used for space weather. Some of the earliest such spacecraft were part of the Applications Technology Satellite (ATS) series at GEO which were precursors to the modern Geostationary Operational Environmental Satellite (GOES) weather satellite and many communication satellites. The ATS spacecraft carried environmental particle sensors as auxiliary payloads and had their navigational magnetic field sensor used for sensing the environment.

Many of the earliest instruments used for monitoring the space environment were and are research spacecraft which were re-purposed or jointly purposed for space weather applications and forecasting. One of the first of these is the IMP-8 (Interplanetary Monitoring Platform) The IMP-8 orbited the Earth at 35 Earth Radii and observed the solar wind for two-thirds of its 12-day orbit from 1973 to 2006. Since the solar wind carries disturbances which affect the magnetosphere and ionosphere, IMP-8 demonstrated the utility of continuously monitoring the solar wind. IMP-8 was followed by ISEE-3 which was placed near the L_1 Sun-Earth Lagrangian point, 235 Earth radii above the surface (about 1.5 million km, or 924,000 miles) and continuously monitored the solar wind from 1978 to 1982. The next spacecraft to monitor the solar wind at the L_1 point was WIND from 1994 to 1998. After April 1998, the WIND spacecraft orbit was change to circle the Earth and pass by the L_1 point occasionally. The NASA Advanced Composition Explorer (ACE) has monitored the solar wind at the L_1 point from 1997 to present. It is estimated to cease operating about 2024. Funding for a replacement for ACE is in the 2012 budget request for NOAA with a planned launch in 2015. The replacement's primary mission will be space weather forecasting and applications.

In addition to monitoring the solar wind, monitoring the Sun is important to space weather. Because the solar EUV cannot be monitored from the ground, the joint NASA-ESA Solar and Heliospheric Observatory (SOHO) spacecraft was launched and has provide EUV images of the Sun from 1995 to the present. SOHO is a main source of near-real time solar data for both research and space weather prediction and inspired the STEREO mission. The Yohkoh spacecraft at LEO observed the Sun from 1991 to 2001 in the X-ray portion of the solar spectrum and was useful for both research and space weather prediction. Data from Yohkoh inspired the Solar X-ray Imager on GOES.

Spacecraft with instruments whose primary purpose is to provide data for space weather predictions and applications include the Geostationary Operational Environmental Satellite (GOES) series of spacecraft, the POES series, the DMSP series, and the Meteosat series. The GOES spacecraft have carried an X-ray sensor (XRS) which measures the flux from the whole solar disk in two bands – 0.05 to 0.4 nm and 0.1 to 0.8 nm – since 1974, an X-ray imager (SXI) since 2004, a magnetometer which measures the distortions of the Earth's magnetic field due to space weather, a whole disk EUV sensor since 2004, and particle sensors (EPS/HEPAD) which measure ions and electrons in the energy range of 50 keV to 500 MeV. Starting sometime after 2015, the GOES-R generation of GOES spacecraft will replace the SXI with a solar EUV image (SUVI) similar to the one on SOHO and STEREO and the particle sensor will be augmented with a component to extend the energy range down to 30 eV.

Space Weather Modelling

Space weather models are computer simulations of the space weather environment. Like computer models for meteorology, space weather models take a limited set of data values and extrapolate to values which describe the entire space weather environment or a segment of the space weather environment in the model. Each model makes a prediction or a set of predictions about how the environment evolves with time. Computer models use the sets of mathematical equations to describe the physical processes involved. The early space weather models were heuristic; *i.e.,* they relate one phenomenon with another without including any physics in the relationship. Some of these simple models are still used because they take minimal resources and yield results which are good enough for some purposes. Present research and development efforts concentrate on complex sets of equations which account for as many elements of physics as possible. Space weather models differ from meteorological model in that amount of input is vastly smaller and no single space weather model yet can reliably predict the environment from the surface of the Sun to the bottom of the Earth's ionosphere.

A significant portion of space weather model research and development in the past two decades has been done as part of the Geospace Environmental Model (GEM) program of the National Science Foundation. Two major centers for modelling are the Center for Space Environment Modelling (CSEM) and the Center for Integrated Space weather Modelling (CISM). The Community Coordinated Modelling Center (CCMC) at the NASA Goddard Space Flight Center is a facility for coordinating the development and testing of research models, for the improvement of models and for preparing models for transition to space weather prediction and application.

Modelling efforts to simulate the environment from the Sun to the Earth use several method including

(a) magnetohydrodynamics in which the environment is treated as a fluid,

(b) *particle in cell* in which non-fluid interactions are handled within a cell and then a series of cells are connected together to describe the environment,

(c) *first principles* in which physical processes are in balance (or equilibrium) with one another,

(d) *semi-static* modelling in which a statistical or empirical relationship is described, or a combination of several of these methods.

Commercial Space Weather Activities

After the start of the space weather discipline in the 1990s, and particularly after 2000, there has been a growing community of commercial space weather providers. These are mostly small companies, that provide a variety of space weather data, models, derivative products, and service distribution for government, commercial, and consumer customers. On April 29, 2010, the commercial space weather community evolved from the Commercial Space Weather Interest Group (CSWIG) that had met for a decade at Space Weather Workshop in Boulder, Colorado every spring, into the American Commercial Space Weather Association (ACSWA). It is a formal industry association representing private-sector commercial interests related to space weather.

Weather Station

A weather station is a facility, either on land or sea, with instruments and equipment for measuring atmospheric conditions to provide information for weather forecasts and to study the weather and climate. The measurements taken include temperature, barometric pressure, humidity, wind speed, wind direction, and precipitation amounts. Wind measurements are taken with as few as other obstructions as possible, while temperature and humidity measurements are kept free from direct solar radiation, or insolation. Manual observations are taken at least once daily, while automated measurements are taken at least once an hour. Weather conditions out at sea are taken by ships and buoys, which measure slightly different meteorological quantities such as sea surface temperature, wave height, and wave period. Drifting weather buoys outnumber their moored versions by a significant amount.

Instruments

Typical weather stations have the following instruments:

- Thermometer for measuring air and sea surface temperature
- Barometer for measuring atmospheric pressure
- Hygrometer for measuring humidity.
- Anemometer for measuring wind speed
- Rain gauge for measuring liquid precipitation over a set period of time.

In addition, at certain Automated airport weather stations, additional instruments may be employed, including:

- Present Weather/Precipitation Identification Sensor for identifying falling precipitation
- Disdrometer for measuring drop size distribution
- Transmissometer for measuring visibility
- Ceilometer for measuring cloud ceiling

More sophisticated stations may also measure the ultraviolet index, solar radiation, leaf wetness, soil moisture, soil temperature, water temperature in ponds, lakes, creeks, or rivers, and occasionally other data.

Exposure

Except for those instruments requiring direct exposure to the elements (anemometer, rain gauge), the instruments should be sheltered in a vented box, usually a Stevenson screen, to keep direct sunlight off the thermometer and wind off the hygrometer. The instrumentation may be specialised to allow for periodic recording otherwise significant manual labour is required for record keeping. Automatic transmission of data, in a format such as METAR, is also desirable as many weather station's data is required for weather forecasting.

Personal Weather Station

Figure: Roof-mounted weather station instruments

A personal weather station is a set of weather measuring instruments operated by a private individual, club, association, or even business (where obtaining and distributing weather data is not a part of the entity's business operation). The quality and number of instruments can vary widely, and placement of the instruments, so important to obtaining accurate, meaningful, and comparable data, can also be very variable.

Today's personal weather stations also typically involve a digital console that provides readouts of the data being collected. These consoles may interface to a personal computer where data can be displayed, stored, and uploaded to Web sites or data ingestion/ distribution systems.

Personal weather stations may be operated solely for the enjoyment and education of the owner, but many personal weather station operators also share their data with others, either by manually compiling data and distributing it, or through use of the internet or amateur radio. The Citizen Weather Observer Program (CWOP) is one such, and the data submitted through use of software, a personal computer, and internet connection (or amateur radio) are utilised by the National Weather Service when generating forecast models, and by many other entities as well. Each weather station submitting data to CWOP will also have an individual Web page that depicts the data submitted by that station. The Weather Underground Internet site is another popular destination for the submittal and sharing of data with others around the world. As with CWOP, each station submitting data to The Weather Underground has a unique Web page displaying their submitted data. The UK Met Office's Weather Observations Website (WOW) also allows such data to be shared and displayed.

Dedicated Ships

Figure: *The weather ship M/S Polarfront at sea.*

A weather ship was a ship stationed in the ocean as a platform for surface and upper air meteorological measurements for use in weather forecasting. It was also meant to aid in search and rescue operations and to support transatlantic flights. The establishment of weather ships proved to be so useful during World War II that the International Civil Aviation Organisation (ICAO) established a global network of 13 weather ships in 1948.

Of the 12 left in operation in 1996, nine were located in the northern Atlantic ocean while three were located in the northern Pacific ocean. The agreement of the weather ships ended in 1990. Weather ship observations proved to be helpful in wind and wave studies, as they did not avoid weather systems like merchant ships tended to and were considered a valuable resource. The last weather ship was MS *Polarfront*, known as weather station M ("Mike") at 66°N, 02°E, run by the Norwegian Meteorological Institute. MS *Polarfront* was removed from service January 1, 2010. Since the 1960s this role has been largely superseded by satellites, long range aircraft and weather buoys. Weather observations from ships continue from thousands of voluntary merchant vessels in routine commercial operation; the Old Weather crowdsourcing project transcribes naval logs from before the era of dedicated ships.

Weather Buoy

Weather buoys are instruments which collect weather and ocean data within the world's oceans, as well as aid during emergency response to chemical spills, legal proceedings, and engineering design. Moored buoys have been in use since 1951, while drifting buoys have been used since 1979.

Moored buoys are connected with the ocean bottom using either chains, nylon, or buoyant polypropylene. With the decline of the weather ship, they have taken a more primary role in measuring conditions over the open seas since the 1970s. During the 1980s and 1990s, a network of buoys in the central and eastern tropical Pacific ocean helped study the El Niño-Southern Oscillation. Moored weather buoys range from 1.5 metres (4.9 ft) to 12 metres (39 ft) in diameter, while drifting buoys are smaller, with diameters of 30 centimetres (12 in) to 40 centimetres (16 in).

Drifting buoys are the dominant form of weather buoy in sheer number, with 1250 located worldwide. Wind data from buoys has smaller error than that from ships. There are differences in the values of sea surface temperature measurements between the two platforms as well, relating to the depth of the measurement and whether or not the water is heated by the ship which measures the quantity.

History

The first known proposal for surface weather observations at sea occurred in connection with aviation in August 1927, when Grover Loening stated that "weather stations along the ocean coupled with the development of the seaplane to have an equally long range, would result in regular ocean flights within ten years." Starting in 1939, United States Coast Guard vessels were being used as weather ships to protect transatlantic air commerce. The Navy Oceanographic Meteorological Automatic Device (NOMAD) buoy's 6-metre (20 ft) hull was originally designed in the 1940s for the United States Navy's offshore data collection program.

Between 1951 and 1970, a total of 21 NOMAD buoys were built and deployed at sea. Since the 1970s, weather buoy use has superseded the role of weather ships by design, as they are cheaper to operate and maintain. The earliest reported use of drifting buoys was to study the behaviour of ocean currents within the Sargasso Sea in 1972 and 1973. Drifting buoys have been used increasingly since 1979, and as of 2005, 1250 drifting buoys roamed the Earth's oceans.

Between 1985 and 1994, an extensive array of moored and drifting buoys was deployed across the equatorial Pacific Ocean designed to help monitor and predict the El Niño phenomenon. Hurricane Katrina capsized a 10 m (33 ft) buoy for the first time in the history of the National Data Buoy Center (NDBC) on August 28, 2005.

On June 13, 2006, drifting buoy 26028 ended its long-term data collection of sea surface temperature after transmitting for 10 years, 4 months, and 16 days, which is the longest known data collection time for any drifting buoy. The first weather buoy in the Southern Ocean was deployed by the Integrated Marine Observing System (IMOS) on March 17, 2010.

Instrumentation

Weather buoys, like other types of weather stations, measure parameters such as air temperature above the ocean surface, wind speed (steady and gusting), barometric pressure, and wind direction. Since they lie in oceans and lakes, they also measure water temperature, wave height, and dominant wave period. Raw data is processed and can be logged on board the buoy and then transmitted via radio, cellular, or satellite communications to meteorological centers for use in weather forecasting and climate study. Both moored buoys and drifting buoys (drifting in the open ocean currents) are used. Fixed buoys measure the water temperature at a depth of 3 metres (9.8 ft). Many different drifting buoys exist around the world that vary in design and the location of reliable temperature sensors varies. These measurements are beamed to satellites for automated and immediate data distribution. Other than their use as a source of meteorological data, their data is used within research programs, emergency response to chemical spills, legal proceedings, and engineering design. Moored weather buoys can also act as a navigational aid, like other types of buoys.

Types

Weather buoys can be recognised by their yellow colour and flashing yellow light at night, and range in diameter from 1.5 metres

(4.9 ft) to 12 metres (39 ft). Those that are placed in shallow waters are smaller in size and moored using only chains, while those in deeper waters use a combination of chains, nylon, and buoyant polypropylene. Discus buoys are round and moored in deep ocean locations, with a diameter of 10 metres (33 ft) to 12 metres (39 ft). The aluminum 3-metre (10 ft) buoy is a very rugged meteorological ocean platform that has long term survivability. The expected service life of the 3-metre (10 ft) platform is in excess of 20 years and properly maintained, these buoys have not been retired due to corrosion. The NOMAD is a unique moored aluminum environmental monitoring buoy designed for deployments in extreme conditions near the coast and across the Great Lakes. NOMADs moored off the Atlantic Canadian coast commonly experience winter storms with maximum wave heights approaching 20 metres (66 ft) into the Gulf of Maine.

Drifting buoys are smaller than their moored counterparts, measuring 30 centimetres (12 in) to 40 centimetres (16 in) in diameter. They are made of plastic or fiberglass, and tend to be either bi-coloured, with white on one half and another colour on the other half of the float, or solidly black or blue. It measures a smaller subset of meteorological variables when compared to its moored counterpart, with a barometer measuring pressure in a tube on its top. They have a thermistor (metallic thermometer) on its base, and an underwater drogue, or sea anchor, located 15 metres (49 ft) below the ocean surface connected with the buoy by a long, thin tether.

Deployment and Maintenance

A large network of coastal buoys near the United States is maintained by the National Data Buoy Center, with deployment and maintenance performed by the United States Coast Guard. For South Africa, the South African Weather Service deploys and retrieves their own buoys, while the Meteorological Service of New Zealand performs the same task for their country. Environment Canada operates and deploys buoys for their country. The Met Office in Great Britain deploys drifting buoys across both the northern and southern Atlantic oceans.

Comparison to Data from Ships

Wind reports from moored buoys have smaller error than those from ships. Complicating the comparison of the two measurements are that NOMAD buoys report winds at a height of 5 metres (16 ft), while ships report winds from a height of 20 metres (66 ft) to 40 metres (130 ft). Sea surface temperature measured in the intake port of large ships have a warm bias of around 0.6 °C (1 °F) due to the heat of the engine room. This bias has led to changes in the perception of global warming since 2000. Fixed buoys measure the water temperature at a depth of 3 metres (10 ft).

Synoptic Weather Station

Synoptic weather station are instruments which collect meteorological information at synoptic time 00h00, 06h00, 12h00, 18h00 (UTC) and at intermediate synoptic hours 03h00, 09h00, 15h00, 21h00 (UTC).

The common instruments of measure are anemometer, wind vane, pressure sensor, air temperature, humidity, and rain-gauge.

The weather measures are formatted in special format and transmit to wmo to help the weather forecast model.

Networks

A variety of land-based weather station networks have been set up globally. Some of these are basic to analyzing weather fronts and pressure systems, such as the synoptic observation network, while others are more regional in nature.

Wind Profiler

A wind profiler is a type of weather observing equipment that uses radar or sound waves (SODAR) to detect the wind speed and direction at various elevations above the ground. Readings are made at each kilometer above sea level, up to the extent of the troposphere (i.e., between 8 and 17 km above mean sea level). Above this level there is inadequate water vapor present to produce a radar "bounce." The data synthesized from wind direction and speed is very useful to meteorological forecasting and timely reporting for flight planning. A twelve hour history of data is available through NOAA websites.

Principle

In a typical implementation, the radar or sodar can sample along each of five beams: one is aimed vertically to measure vertical velocity, and four are tilted off vertical and oriented orthogonal to one another to measure the horizontal components of the air's motion. A profiler's ability to measure winds is based on the assumption that the turbulent eddies that induce scattering are carried along by the mean wind. The energy scattered by these eddies and received by the profiler is orders of magnitude smaller than the energy transmitted. However, if sufficient samples can be obtained, then the amplitude of the energy scattered by these eddies can be clearly identified above the background noise level, then the mean wind speed and direction within the volume being sampled can be determined. The radial components measured by the tilted beams are the vector sum of the horizontal motion of the air toward or away from the radar and any vertical motion present in the beam. Using appropriate trigonometry, the three-dimensional meteorological velocity components (u,v,w) and wind speed and wind direction are calculated from the radial velocities with corrections for vertical motions.

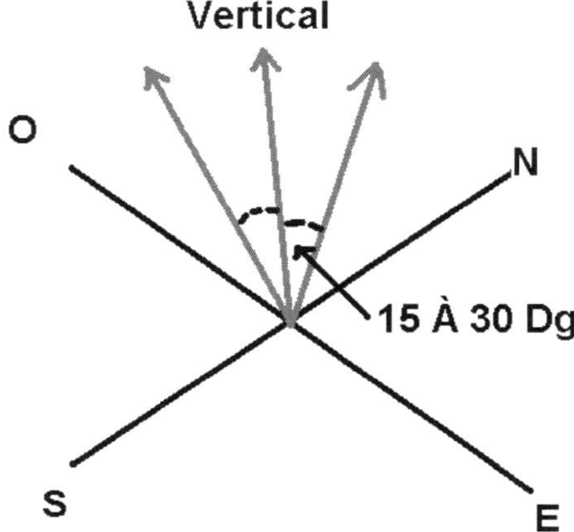

Figure: *Orientation of the beams in the case of a three tilted wind profiler*

Radar Wind Profiler

Figure: *A radar wind profiler.*

Pulse-Doppler radar wind profilers operate using electromagnetic (EM) signals to remotely sense winds aloft. The radar transmits an electromagnetic pulse along each of the antenna's pointing directions. A UHF profiler includes subsystems to control the radar's transmitter, receiver, signal processing, and Radio Acoustic Sounding System (RASS), if provided, as well as data telemetry and remote control.

The duration of the transmission determines the length of the pulse emitted by the antenna, which in turn corresponds to the volume of air illuminated (in electrical terms) by the radar beam. Small amounts of the transmitted energy are scattered back (referred to as backscattering) toward and received by the radar. Delays of fixed intervals are built into the data processing system so that the radar receives scattered energy from discrete altitudes, referred to as range gates. The Doppler frequency shift of the backscattered energy is determined, and then used to calculate the velocity of the air toward or away from the radar along each beam as a function of altitude. The source of the backscattered energy (radar "targets") is small-scale turbulent fluctuations that induce irregularities in the radio refractive index of the atmosphere. The radar is most sensitive to scattering by turbulent eddies whose spatial scale is ½ the wavelength of the radar, or approximately 16 centimeters (cm) for a UHF profiler.

A boundary-layer radar wind profiler can be configured to compute averaged wind profiles for periods ranging from a few minutes to an hour. Boundary-layer radar wind profilers are often configured to sample in more than one mode. For example, in a "low mode," the pulse of energy transmitted by the profiler may be 60 m in length. The pulse length determines the depth of the column of air being sampled and thus the vertical resolution of the data. In a "high mode," the pulse length is increased, usually to 100 m or greater. The longer pulse length means that more energy is being transmitted for each sample, which improves the signal-to-noise ratio (SNR) of the data. Using a longer pulse length increases the depth of the sample volume and thus decreases the vertical resolution in the data. The greater energy output of the high mode increases the maximum altitude to which the radar wind profiler can sample, but at the expense of coarser vertical resolution and an increase in the altitude at which the first winds are measured. When radar wind profilers are operated in multiple modes, the data are often combined into a single overlapping data set to simplify post-processing and data validation procedures.

Sodar Wind Profiler

Alternatively, a wind profiler may use sound waves to measure wind speed at various heights above the ground, and the thermodynamic structure of the lower layer of the atmosphere. These sodars can be divided in mono-static system using the same antenna for transmitting and receiving, and bi-static system using separate antennas. The difference between the two antenna systems determines whether atmospheric scattering is by temperature fluctuations (in mono-static systems), or by both temperature and wind velocity fluctuations (in bi-static systems).

Mono-static antenna systems can be divided further into two categories: those using multiple axis, individual antennas and those using a single phased array antenna. The multiple-axis systems generally use three individual antennas aimed in specific directions to steer the acoustic beam. One antenna is generally aimed vertically, and the other two are tilted slightly from the vertical at an orthogonal angle. Each of the individual antennas may use a single transducer focused into a parabolic reflector to form a parabolic loudspeaker, or an array of speaker drivers and horns (transducers) all transmitting in-phase to form a single beam. Both the tilt angle from the vertical and the azimuth angle of each antenna are fixed when the system is set up.

The vertical range of sodars is approximately 0.2 to 2 kilometers (km) and is a function of frequency, power output, atmospheric stability, turbulence, and, most importantly, the noise environment in which a sodar is operated. Operating frequencies range from less than 1000 Hz to over 4000 Hz, with power levels up to several hundred watts. Due to the attenuation characteristics of the atmosphere, high power, lower frequency sodars will generally produce greater height coverage. Some sodars can be operated in different modes to better match vertical resolution and range to the application. This is accomplished through a relaxation between pulse length and maximum altitude.

Bibliography

Aczel, Amir D.: *The Riddle of the Compass: The Invention that Changed the World*, Harcourt, New York, 2001.

Bothmer, V.; Daglis, I.: *Space Weather: Physics and Effects*, Springer-Verlag, New York, 2006.

Campbell, Wallace H.: *Introduction to geomagnetic fields*, Cambridge University Press, New York, 2003.

Carlowicz, M. J., and R. E. Lopez: *Storms from the Sun*, Joseph Henry Press, Washington DC, 2002.

Comins, Neil F.: *Discovering the Essential Universe*, W. H. Freeman, New York, 2008.

Daglis, I. A.: *Space Storms and Space Weather Hazards*, Springer-Verlag, New York, 2001.

Feynman, Richard P.: *The Feynman lectures on physics,* BasicBooks, New York, 2010.

Freeman, John W.: *Storms in Space*, Cambridge University Press, Cambridge, UK, 2001.

Gies, Frances and Gies, Joseph: *Cathedral, Forge, and Waterwheel: Technology and Invention in the Middle Age*, HarperCollins, New York, 1994.

Gombosi, Tamas I., Houghton, John T., and Dessler, Alexander J.: *Physics of the Space Environment,* Cambridge University Press, 2006.

Ioannis A. Daglis: *Effects of Space Weather on Technology Infrastructure*, Springer, Dordrecht, 2005.

Jean Lilensten and Jean Bornarel: *Space Weather, Environment and Societies*, Springer, New York, 2009.

Lorimer, Duncan R.; Kramer, Michael: *Handbook of Pulsar Astronomy, Cambridge University Press*, 2004.

Lyne, Andrew G.; Graham-Smith, Francis: *Pulsar Astronomy*, Cambridge University Press, 1998.

Mark, Moldwin: *An introduction to space weather*, Cambridge Univ. Press, Cambridge, 2008.

McElhinny, Michael W.; McFadden, Phillip L.: *Paleomagnetism: Continents and Oceans*, Academic Press, New York, 2000

Merrill, Ronald T.: *Our Magnetic Earth: The Science of Geomagnetism*, The University of Chicago Press, Chicago, 2010.

National Research Council: *Plasma Processing of Materials : Scientific Opportunities and Technological Challenges*, National Academies Press, New York, 1991.

Needham, Joseph: *Science and civilisation in China*, Caves Books, originally publ. by Cambridge University Press, Taipei, 1986.

Odenwald, S.: *The 23rd Cycle; Learning to live with a stormy star*, Columbia University Press, 2006.

Opdyke, Neil D.; Channell, James E. T.: *Magnetic Stratigraphy*, Academic Press, 1996.

Parks, George K.: *Physics of space plasmas: an introduction*, Addison-Wesley, New York, 1991.

R. Zubrin: *Entering Space: Creating a Spacefaring Civilization*, Jeremy P. Tarcher/Putnam, New York, 1999.

Radu Balescu : *Aspects of Anomalous Transport in Plasmas*, CRC Press, United States, 2005 .

Raine, Derek J. and Thomas, Edwin George: *Black Holes: An Introduction*, Imperial College Press, London, 2010.

Shalom, Eliezer: *The Fourth State of Matter: An Introduction to Plasma Science*, CRC Press, US., 2001 .

Song, P., Singer, H., and Siscoe, G.: *Space Weather,* Union, Washington, D.C., 2001.

Tauxe, Lisa: *Paleomagnetic Principles and Practice*, Kluwer, Netherlands, 1998.

Vacquier, Victor: *Geomagnetism in marine geology,* Elsevier Science, Amsterdam, 1972.

Von Engel, A. and Cozens, J.R.: *"Flame Plasma" in Advances in electronics and electron physics*, Academic Press, US., 1976.

Walt, Martin: *Introduction to Geomagnetically Trapped Radiation*, Cambridge University Press, 1994.

Williams, J.E.D.: *From Sails to Satellites: the origin and development of navigational science*, Oxford University Press, 1992.

Yaffa Eliezer, Shalom Eliezer: *The Fourth State of Matter: An Introduction to the Physics of Plasma*, Adam Hilger, London, 1989.

Zhou, Daguan: *The customs of Cambodia*, Indochina Books, Phnom Penh, 2007.

Index

I

K

L

M

N

O

P

R

S

T

U

V

W

❑❑❑